WESTMAR COLLE[illegible]
W9-APW-815

Ecology in Ancient Civilizations

J. Donald Hughes

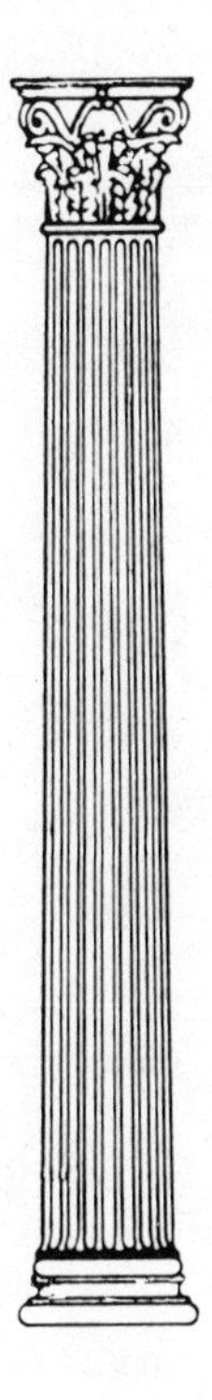

UNIVERSITY OF NEW MEXICO PRESS
Albuquerque

301
.31093
H893
GF
541
.H83

© The University of New Mexico Press, 1975. All rights reserved.
Manufactured in the United States of America. Library of Congress Catalog Card Number 74-27446. International Standard Book Number 0-8263-0367-6. First edition.

92348

TO PAMELA

ὅσῳ πλέον ἥμισυ παντὸς . . .
. . . ὅσον ἐν μαλάχῃ τε καὶ ἀσφοδέλῳ μέγ' ὄνειαρ.

How much better it is to share
Than to keep all for oneself . . .
How much refreshment of life there is
In the beauty of mallow and asphodel.
—Hesiod, *Works and Days*

Preface

This book is intended to serve as an introduction to the environmental history of the ancient world, and is written for the general reader and student rather than the specialist. I believe the subject is interesting not only in itself, but also as a means of understanding the early antecedents of the modern ecological crisis. Thus I have given attention chiefly to the more direct cultural ancestors of today's technological society: the ancient peoples of the Mediterranean Basin, in particular the Greeks and Romans, the Jews, and the early Christians.

Environmental history is still a new and challenging field. The researcher must read widely and combine insights from several areas of inquiry which are usually kept separate. I hope this book may help to stimulate further investigation and writing by others.

I am indebted to friends and colleagues for encouragement and assistance at various stages in the preparation of this book, and wish to express thanks, especially to some whose help was indispensable. The subject was suggested to me by Ray Gilmore of the San Diego Museum of Natural History. A constant friend, Allen D. Breck, chairman of the Department of History at my university, has given my work support and understanding. Wolfgang Yourgrau, historian and philosopher of science, has faithfully fulfilled the office of senior colleague, providing time and advice generously. Departmental lines have not been barriers for friends who have shared the results of their own research and writing with me, especially Moras Shubert, biologist and ecologist; Charles Stevens, geographer and environmentalist; and Cynthia Cohen, philosopher.

The University of Denver supported my research and gave me a sabbatical quarter, which enabled me to complete this book. For those of my students whose indefatigable interest has opened up new lines of inquiry for me—John Chase, Alexandra Hendee, Erich Wiesinger, and others—I treasure lasting esteem. Special thanks is due to colleagues at other universities who have given me forums in which to advance and test some of the ideas contained herein, particularly Robert Hohlfelder of the University of Colorado, Gerald Else of the University of Michigan, and also the editorial staff of *Inquiry: An Interdisciplinary Journal of Philosophy and the Social Sciences,* where an article of mine on a related subject will soon appear.

The respected ecologist, Paul B. Sears, read my manuscript when it was near completion and made several valuable suggestions, for which I am grateful.

Contents

MAPS AND ILLUSTRATIONS

Maps

Illustrations

following page 82

Color Plates

following page 114

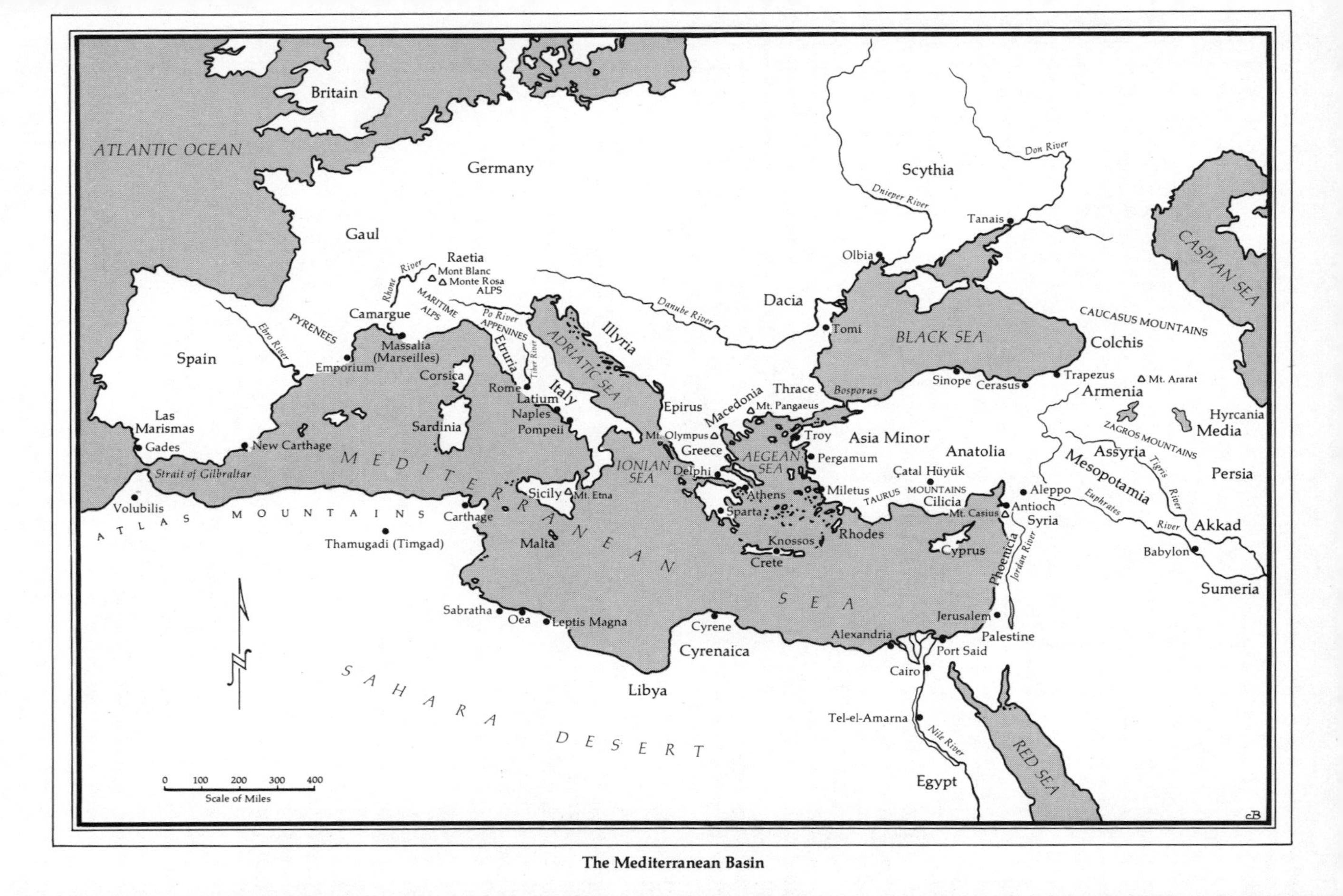

The Mediterranean Basin

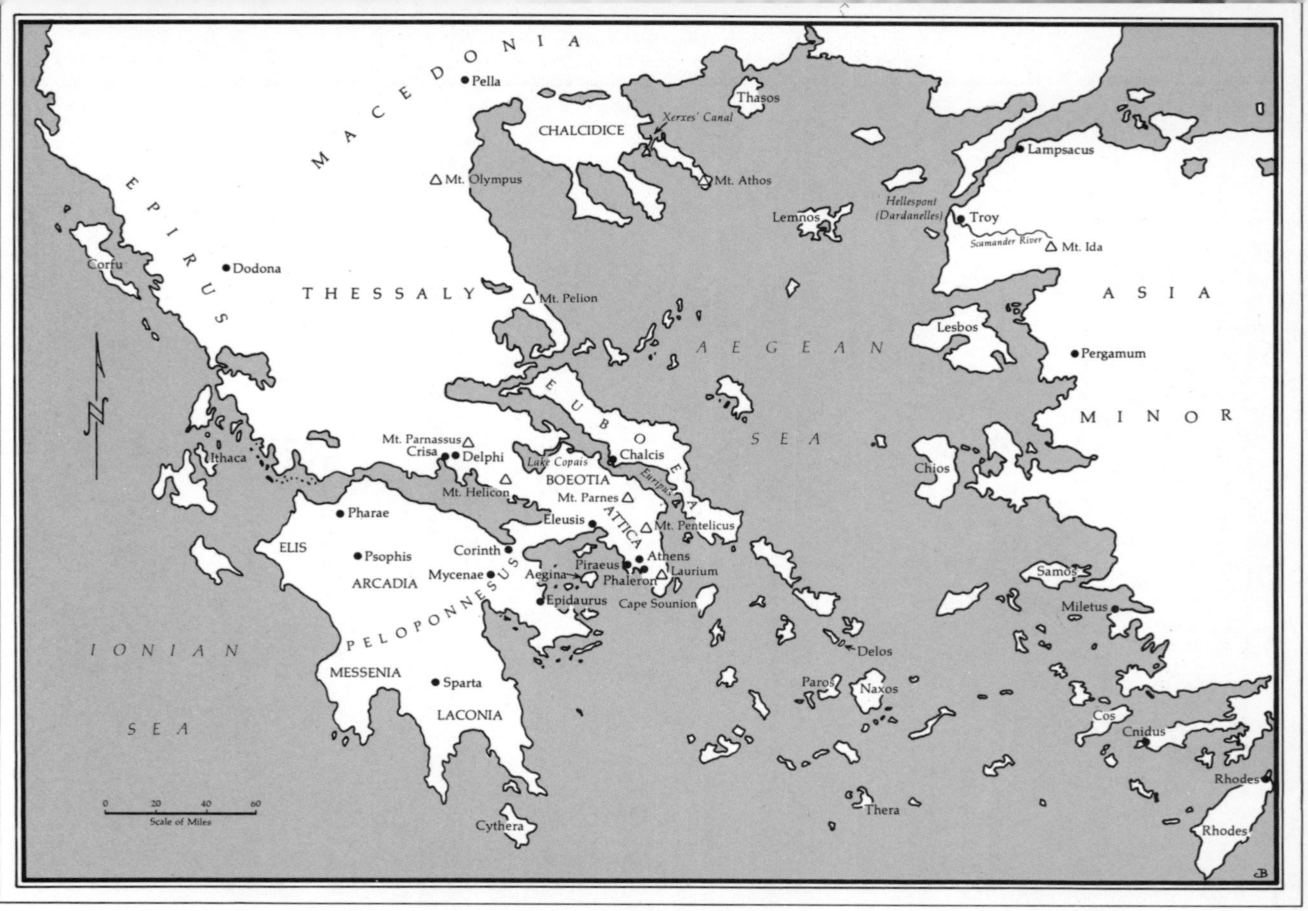

Greece and the Aegean

A L P S
Po River
A P P E N N I N E S
ADRIATIC SEA
Ravenna
Etruria
Tiber River
Ciminian Forest
Corsica
Alalia
Rome
Aricia
Ostia
Ardea
Latium
Pontine Marshes
Fount of Bandusia
Naples
Mt. Vesuvius
Pompeii
BAY OF NAPLES
Siris River
Paestum
Sardinia
Thurii
TYRRHENIAN SEA
Panormus (Palermo)
Mt. Etna
Sicily
Syracuse
0
50
100
150
200
Scale of Miles

Italy

1

Environment and Civilization

Those who look at the Parthenon, that incomparable symbol of the achievements of an ancient civilization, often do not see its wider setting. Behind the Acropolis, the bare dry mountains of Attica show their rocky bones against the blue Mediterranean sky, and the ruin of the finest temple built by the ancient Greeks is surrounded by the far vaster ruins of an environment which they desolated at the same time. In the centuries before the Golden Age of Athens, those same mountains were covered by forests and watered by springs and streams. The philosopher Plato saw evidence of the changes that had occurred not long before; there were buildings in Athens with beams fashioned from trees that had grown on hillsides which by his day were eroded and covered only with herbs, and he visited shrines once dedicated to the guardian spirits of flowing springs which had since dried up.[1]

The huge Roman cities of North Africa have only recently been excavated from the drifted sands of the Sahara Desert which had covered them for centuries. Now the long, wide streets, the theaters and marketplaces stand uninhabited in a sterile landscape, but in ancient times they supported large populations and exported wheat, olive oil, and other agricultural products to Rome. The decline and fall of the Roman Empire evidently had an environmental dimension.

The cedar forests of Lebanon supplied the best wood not only for the construction of Solomon's Temple in Jerusalem, but also for numerous cities of the Near East. Today a few tiny, protected groves manage to survive among whole mountain ranges of dry, eroded rock. The once-prosperous capital cities of Mesopotamia are now mounds of clay in the desert, where the courses of dry canals may be traced under dust and sand blown there by the winds. The famous "Fertile Crescent" of early agriculture once arched from the borders of Egypt through Syria and Mesopotamia to the Persian Gulf, but pictures taken from space by the astronauts show that it is now a shrunken remnant.

Throughout the Mediterranean Basin and the adjoining Near East, the ruins of ancient civilizations stand amid the evidences of depleted environment. The conclusion seems inescapable that the natural environment and the course of civilizations were interrelated. It is the purpose of this book to study that interrelationship from the viewpoint of environmental history, which is an attempt to apply the insights of ecology to human history. An effort such as this requires digging into the records of ancient civilizations, studying both their writings and their archaeological remains. For the author, it requires familiarity with ancient history and environmental science.

It is a subject on the borderline between two vast fields, but an ecological view of ancient civilizations in that part of the world where modern Western society found its roots promises to be rewarding, since it may reveal insights not only into the fates of those vanished civilizations but also into the development of human relationships with and attitudes toward the natural environment down to the present day.

Ecology, as understood in this regard, is the study of the interrelationships of living things to one another and their surrounding environment. Thus considered, it is a branch of biology, the study of living things. Ecology is a wide-ranging study of whole assemblages of living things as they interact with one another. These interacting assemblages are called *ecosystems.*

Ecology avoids concentrating too long on any one animal or plant in isolation, and tries to see that everything is connected to everything else. The human species is no exception to this last statement, and in a discussion of ecology and ancient civilizations, interest centers on mankind.

The very word *ecology,* in its Greek roots, reflects an early concern with humanity. *Oikos* means *home* or *house* in Greek, and by extension it came to mean the whole inhabited earth, the *oikoumene,* the home of all mankind. *Logos,* meaning *reason* or *study,* is a common suffix applied to many of the sciences, indicating the human mind at work on a given subject. Human ecology, then, is a rational study of how mankind interrelates with the home of the human species, the earth; with its soil and mineral resources; with its water, both fresh and salt; with its air, climates, and weather; with its many living things, animals and plants, from the simplest to the most complex; and with the energy received ultimately from the sun.

In the historical study of the relationship of human civilizations to the natural environment, certain themes present themselves, and among these, three seem basic and central: first, the influence of the environment on the development of civilizations; second, human attitudes toward nature; and third, the impact of civilizations upon the natural environment. Of these, the present study will emphasize the second and third themes.

The natural environment had a constant, formative influence on ancient civilizations. This first theme has been considered by many writers, beginning with Hippocrates, the Greek father of medicine, who noticed the effect of climate on human health, temperament, and intelligence and remarked that civilizations arose in lands of moderate or warm climate with light rainfall, where water supply was a major challenge.

One of the most striking contrasts between the ancient and modern worlds is the more direct dependence on the natural world which ancient peoples both had and felt. There was no vast technology then interposing between human beings and nature. Intimacy with nature and sensitivity to its cycles was the rule, not the exception.

Of course, the natural environment is not the only influence on developing civilizations; cultural traits, the past experience of peoples, mobility, and the level of technological development are important, too. But it is a major factor. The Greeks and Phoenicians, with ready access to the sea, were much more likely to develop naval and merchant skills than the landlocked Assyrians and Persians. The types of soil and terrain available to a society placed limits on the nature and extent of the food supply. The thin, hilly soil of Attica, for example, was suited to growing olive trees and grapevines but not to extensive grain cultivation. The presence or absence of various minerals and stone for quarrying was important; the Egyptians erected stone monuments, but the Mesopotamians in their alluvial valley built with clay bricks. The manufacture of bronze depended on tin, and some ancient nations had to import tin from mines hundreds of miles away.

The kinds of animals and plants in the environment affected civilizations, too. Some could be domesticated or cultivated and introduced to other areas where the climate was favorable to them. The Greeks introduced the olive almost everywhere they settled.

The art and literature of all civilizations are full of motifs inspired by the natural world. Examples such as the Minoan frescoes of octopuses and leaping dolphins, the Assyrian reliefs of lions, or Homer's description of the forest near the Cave of Calypso in the *Odyssey* occur to mind. Philosophy was affected, too; Thales, who said the world was made of water, lived on the island-dotted shore of the Aegean Sea, not in the arid interior of Asia. Religion was deeply imbued with nature. Most early religions emphasized fertility and honored certain animals, plants, and locations; Christians use wine and bread in the sacrament because these were the staple food and drink of the land of Christianity's birth. This list of influences of the natural environment on civilizations is far from complete. The theme of environmental influence on civilization is vast and pervading, and has been the subject of many prior studies. In a book such as this one, some comment upon it is unavoidable. But the other two themes mentioned above, which emphasize human response to and effects on the natural environment, have received considerably less attention and will be the major concerns here.

The relationship of these ancient civilizations to the natural environment was determined in part by their characteristic attitudes toward nature. Certainly the actions of a people tend to reflect their perceptions and their values. A people who, for the most part, regard certain natural objects as sacred will treat them in a different way from those who regard them merely as things to use. Modern attitudes find precedents in the ancient world; if some modern environmental problems are similar to those faced by ancient peoples, some of them may also be due to the perpetuation of ancient attitudes, attitudes varying from worship and curiosity to domination and use. In this book, considerable attention will be given to the question, What were the attitudes of ancient peoples toward nature?

Human civilizations have altered the environments in which they developed. Some of the effects have been advantageous for mankind and have made it possible for people to live in a symbiotic balance with nature. The examples which began this chapter, however, the deforestation of Greece and Lebanon and the invasion of Roman and Mesopotamian cities by the desert, are evidence of human mistreatment of nature, and the conclusion seems inescapable that nature has had her revenge in the fall of civilizations. The dichotomy of human activities and the natural environment is false, even though it is useful for the purpose of discussion, since mankind is part of nature and both acts upon and is acted upon by the rest of the natural world. But to a greater extent than any other creature, mankind has shown an ability to alter, shape, and interfere in the interrelationships of all creatures. This ability and its results are a major cause of the successes and failures of ancient civilizations. As Ellen Semple wisely observed,

> The causes of decline are to be sought . . . in denudation of the hillside soil, deforestation with the failure of springs, destruction of irrigation works by barbarian or nomad attack, collapse of orderly government under repeated barbarian inroads, and possibly to exhaustion of the soil, causing agricultural decline.[2]

It is this aspect of the relationship between civilization and the environment, that is, the impact of human activities on the natural world, along with the human attitudes which help to shape those activities, which will be the major subject of the balance of this book. How did civilizations alter the natural environment, and why?

This book is intended as a provocative introduction to a broad subject which impinges on many fields of knowledge. Environmental history is a rapidly expanding discipline, and it is the author's hope that new investigations will add to our knowledge of environment and mankind in the ancient period, and thus help us in our understanding of our present environmental crisis, a subject to which we will return in the last chapter.

2

The Mediterranean Ecosystem

The ecosystem of the Mediterranean Basin and the adjoining river valleys, within which the ancient civilizations of the West arose, is varied. It can be harsh to those who live there, and even violent at times, but all things considered, it is one of the most pleasant areas of the earth. It is rich and splendid enough to enchant people even today, and many an American or Northern European has found in it what Odysseus's men found in the "land of the lotus eaters,"[1] which after all was itself a Mediterranean island, a place too bountiful and seductive to leave. It is also an area which has felt the results of human actions as long and as extensively as any place on earth. The many splendors of the Mediterranean natural environment as it was before being changed by civilized humanity can only be imagined, or glimpsed in the few relatively undisturbed spots which remain, such as Mount Athos or parts of the French Camargue.

The geographic area described here comprises the theater of ancient Western civilization, the Mediterranean Sea and the lands bordering it and draining into it; the Black Sea, which is "a vast alcove of the Mediterranean,"[2] and its immediate coastlands; and the adjacent river valleys of the Tigris, Euphrates, and Jordan. To call such an area an ecosystem is a convenience. In the largest sense, the earth itself is an ecosystem, and in another sense, the Mediterranean area is made up of many communities of animals and plants, each of which could be considered, with its physical setting, to be an ecosystem.

The Mediterranean is the largest inland sea in the world, surrounded almost completely by the land masses of Europe, Asia, and Africa, and penetrating deeply into them. Its only connection with the rest of the world's oceans is through the Strait of Gibraltar, called the Pillars of Heracles by the Greeks and the Pillars of Melkart by the Phoenicians, only nine miles wide and a maximum of twelve hundred feet deep. This shallow, narrow channel provides a huge supply of inflowing water, because the Mediterranean is at a lower average level than the Atlantic here. The Mediterranean loses more water by evaporation than it gains from rain and the rivers that enter it, including the Ebro of Spain, the Rhone of France, the Po of Italy, and even the Nile of Egypt, one of the great rivers of the earth. The water that enters from the Atlantic is relatively warm; there is no way for the colder waters of the depths to get in.

The Black Sea is formed like a miniature Mediterranean, separated from the larger sea by even narrower passages, the Dardanelles (Hellespont) and Bosporus, whose narrowest point is only eight hundred yards wide. Here the flow is also into the Mediterranean. A current running about three miles an hour pours out of the Black Sea, which is much smaller and lies in cooler northern latitudes, thus suffering less evaporation, and receives the ample waters of the Danube, Dnieper, Don, and other rivers. The salt content of the Black Sea, at least in its upper levels, is comparatively low, whereas that of the Mediterranean is exceptionally high.

The Mediterranean is an almost tideless sea. Enclosed as it is, it cannot reflect the influence of the moon's gravitational pull by anything more than a little sloshing about in its own basin. Tides in most places around the Mediterranean are less than three feet between high and low, and in Port Said and Marseilles, for example, they are less than one foot. (For comparison, New York has a six-foot tide, Boston's is twelve feet, and Seattle's thirteen and a half feet.) As a result, building and cultivation can often be undertaken safely next to the sea. The Mediterranean does not usually have much of a surf, although storms can build up the waves rapidly, especially in winter, and from time to time tidal waves are caused by earthquakes and volcanic disturbances.

Today, almost all the ancient ports of the Mediterranean coast are at least partly under water. The water of this sea is extremely clear most of the time, and the patterns of ancient docks and breakwaters can be traced under the sea some distance out beyond their modern counterparts. Although local vertical movements of the earth's crust do occur, particularly in areas of volcanic activity, it is evident that the Mediterranean Sea is deeper today than it was in classical times. This is probably the result of the melting of polar ice in a general warming trend, rather than subsidence of the Mediterranean Basin. The Strait of Gibraltar would be no barrier in such a long-term process. There is evidence to indicate that the level of the inland sea has risen and fallen in the past. The level of the Mediterranean, which is less subject to tides than other seas, is a good indicator of the overall temperature of the earth.

The Mediterranean climate is even, dependable, and quite distinct. It may be described briefly as a climate with two seasons: a hot, dry summer and a mild, moist winter. While similar zones occur on the southwestern edges of all the other continents—in California, central Chile, the Cape of Good Hope area of South Africa, and south and western Australia—the unique position and conformation of the Mediterranean Basin, with the sea extending twenty-five hundred miles into the land, have made possible by far the greatest extent of this climate.

The wet season lasts from October to April, longer in the north and west, and shorter in the south and east. Practically all the annual rainfall, which averages between fifteen and thirty-five inches per year, falls in this season. The "wet season" is not one of constant rain; rainstorms are intense but infrequent in most places, and much of the winter is sunny. Rainfall is the most unpredictable element in the climate; one year may provide twice the average, another less than half. Mountain ranges intercept the prevailing moist winter winds from the west, so that rainfall is much higher on the west-facing slopes, and much lower in "rain shadows" to the east of the mountains. For example, Corfu, a hilly island on the western edge of Greece, receives forty-eight inches of rain annually, while Athens, east of the Greek mountains, receives only sixteen. It is definitely not true that "the rain in Spain stays mainly in the plain." In most of the Mediterranean area, plant growth is more limited by lack of rain than by winter cold.

Temperatures are moderated by the nearness of the sea, and rapid changes seldom occur. Average temperatures in winter are mild; in Palermo, Sicily, for example, there is no month with an average temperature below fifty degrees, and in Athens one rarely needs an overcoat. The daytime and nighttime temperatures may differ by fifteen degrees or so. Nights with frost are unusual near sea level, and when snow falls it is a brief novelty. In the mountains, however, winters are colder. Temperatures fall rapidly with elevation, and the higher mountain ranges of Greece and Italy are snowcapped throughout the winter. The shepherds of the Mediterranean highlands wrap themselves in thick goatskin cloaks with good reason.

Winter was not the season for navigation in ancient times. Storms lash the sea into high waves, and shipwrecks are common even today. The proper season for sailing is the summer, with its moderate and dependable winds.

The dry, hot season lasts from April to October, and the prevailing wind is the steady "Etesian" from the northeast. A few violent summer thunderstorms may sweep out from the mountains and send flash floods rolling down otherwise dry riverbeds, but generally there is no rain.

The summer heat is also moderated by the sea, and temperatures on the average are only about twenty-five degrees warmer than in winter. But heat of eighty or ninety degrees in a maritime climate can be very oppressive, and the Mediterranean peoples generally, not just the Spaniards, take siestas in the summertime.

The Mediterranean light, particularly during the summer, is noted for its clarity and brilliance. Some scholars have gone so far as to suggest that the special quality of this light had much to do with the development of Greek architecture, with its fluted columns and attention to lines and shadows, and they may be partly right. But it seems undeniable that the Mediterranean climate encouraged outdoor activities of all kinds, and goes far to explain the feeling of Mediterranean people for nature.

It seems natural in speaking of the Mediterranean area to discuss the sea first, and then the land, as I have done here. The land in the Mediterranean Basin is marked by its intimate association with the sea. In countless islands, long peninsulas and headlands, bays and deep inlets, the land and the sea intermix, and there are few places on the land very far from the sea. The exceptions are the mountain-ringed plateaus of Anatolia and Spain, and Upper Egypt and Mesopotamia, where the rivers, not the sea, dominate.

The land is characteristically mountainous. The mountains are rugged and extensive, and although they seldom exceed ten or eleven thousand feet in height, they are bold and beautiful, rising as they so often do near the sea. Mount Athos in Greece towers sixty-four hundred feet directly above the Aegean and is the site of a famous shipwreck of the Persian navy. Both Mount Etna and Mount Olympus are nearly ten thousand feet high, and are seen by ships far out at sea. The barrier range of the Alps, which divides the Mediterranean Basin from the rest of Europe, has several peaks, including Mont Blanc and Monte Rosa, higher than any in the forty-eight contiguous United States. But the mountains of the Mediterranean area are more notable for their number and extent than for the elevation of a few spectacular peaks.

There are two exceptions to the general rule that Mediterranean lands are rugged and mountainous: the level wheat-bearing plains of Scythia (Southern Ukraine), which are touched by the Black Sea, and the long, low desert coast of Egypt and Libya, broken only by the highland of Cyrenaica. The plains in the rest of the basin are alluvial, deposited by rivers in valleys between the mountains.

The Mediterranean Basin is a geologically active area, where three great plates of the earth's crust come together. Earthquakes are often disastrous and can be expected from year to year. No Mediterranean locality is entirely free from them. There are also many volcanoes. A few, like Etna and Vesuvius, are considered active, since they erupt every so often. Others are quiescent. They should probably not be called extinct, since some, like the Aegean island volcano Thera, have been known to have great eruptions at intervals of thousands of years. During periods of volcanic activity, the winds carry ash and dust for many miles, enriching the soil of whole districts. The finest dust is carried all over the earth, and in the months that go by before it settles, it adds unusual color to sunrises and sunsets and screens out some of the sun's rays before they reach the surface of the earth.

An amazing variety of living things flourish in the Mediterranean ecosystem. Various parts of the basin support different communities of animals and plants, interacting with one another, each adapted to its particular location and climate. The number of species just of flowering plants is staggering; Greece alone, with only half the area of the British Isles, has three times as many kinds of flowering plants, well over 6,000 as compared to Britain's 2,113. The most important plant communities will be described here, but it is impossible even to mention all of the more important species. Instead, a few plants typical of each community will be listed.

In much of the area, there is a remarkable growth of several kinds of large shrubs, called *maquis* in French and *macchia* in Italian. There is no English name for this vegetative community, but in California the Mediterranean-type brushlands are called "chaparral," from the Spanish term for the scrub oak, *chaparro*. The stiff, thick character of this unique vegetation can be gathered from the fact that cowboy "chaps" were designed of leather to protect the legs of riders against its sharp branches and foliage. Maquis is the dominant climax vegetation around much of the Mediterranean, and is not always simply the result of deforestation. Although maquis never grows much over twenty feet tall, and is more usually twelve to fifteen feet tall in its natural state, it is considered a forest type in this climate zone. Trees such as evergreen oaks, pines, and juniper often grow in association with the maquis, and plane trees mark permanent sources of water. But the dominant plants in this community are large shrubs such as kermes oak, arbutus, laurel, myrtle, tree heather, rockrose, broom, mastic tree, and rosemary. All of them are adapted to the long, dry summer, with its constant danger of fire, and to the mild winter. They are evergreens, taking advantage of the winter moisture and the growth that is possible then. Their leaves are usually dark green, covered by a thick, shiny, waxy or leathery layer to retard evaporation and conserve water in the summertime. They bloom during the short spring, when the weather is briefly both warm and moist, producing a burst of color and fragrance. Most of the species regenerate rapidly after a fire, either growing up again from the root crown or having seeds that germinate in great heat or spread easily and germinate on bare ground. Although maquis may appear to be useless brushland to those accustomed to well-watered and thickly forested lands, people who live in the Mediterranean climate find it very useful as the chief source of firewood and charcoal, pasture for goats, and a place to gather tannin, dyes, gums, resins, and fibers. Of course, maquis cannot stand the constant pressure of destruction and removal. A hillside where the maquis has been stripped away will suffer extreme erosion in the brief, torrential rains of winter. If regeneration does not take place, because of goats browsing heavily, repeated burning, or clearing for cultivation, the soil will virtually disappear so that the earth's rocky skeleton shows through, and a lower, more tenacious vegetative community called *garigue* will appear.

Garigue, or "rock heath," is widespread on rocky hillsides and sandy or gravelly tracts. It is a natural part of the Mediterranean scene, but there is no doubt that man and his animals have produced far more of it than originally existed. The Greeks call it *phrygana,* a word which has also come to mean "kindling," from the thin sticks that are all the wood it supplies. In Spain it is called *tomillares,* after the thyme which grows in it; and it is the source of many other aromatic herbs and spices such as oregano, rosemary, basil, sage, savory, rue, hyssop, lavender, and garlic. Returning sailors in the Mediterranean often catch the pleasant smell of their homeland while they are still far out at sea. Garigue consists of scattered green tufts of "subshrubs" among the rocks, rarely growing more than a foot or two tall. In Greece it comprises over two hundred different species of plants, all of which are adapted to extreme dryness and exposure. Most of them have very small, leathery leaves, often covered with a bloom of whitish hairs to retard evaporation, and thorns to discourage grazing animals. The only tree of the garigue is the dwarf palm of the western Mediterranean. Low junipers, broom, gorse, spurges, rockroses, and daphne, as well as the aromatic plants mentioned above, are typical members of the garigue. Also common are bulbous and tuberous plants like asphodels, irises, tulips, crocuses, and grape hyacinths. Like the maquis, the garigue has a brief, showy explosion of flowers in the spring.

In even drier areas such as the Libyan coast, the valley between the Atlas Ranges, central Spain, and the heart of Asia Minor, or in locations excessively exploited by man, not even the woody plants of the garigue can survive, and there is instead a dry subtropical "steppe." This winter grassland should not be confused with the great temperate plains, also called "steppe," which are summer grasslands of dry continental areas to the north, such as southern Russia. The Mediterranean steppe is green in the moist half of the year. It contains many annual flowers and root perennials, as well as bulbous and tuberous plants which produce brief spring riots of color and then wither before the hot, dry blasts of summer. "The grass withers, the flower fades."[3]

Even at lower elevations, the ancient Mediterranean did not lack true forests, although all but a few remnants have now disappeared. The most important forest trees are the evergreen oaks, especially the holm oak, and the pines, particularly the widespread and drought-resistant Aleppo pine. Cypress, cedar, laurel, carob, and wild olive also formed forests. Most of these forests are open in nature, with trees spaced apart to catch the available moisture in their root systems.

Mediterranean forests show a typical zonation according to altitude above sea level, and form "life zones" something like those in the western United States. The evergreen forest zone, which includes the trees described above, as well as the maquis and garigue, extends up to fifteen hundred or three thousand feet or more in elevation. Above that level, where there is sufficient moisture, and particularly on the western slopes of mountain ranges, the deciduous forest zone occurs, with deciduous oaks, elm, beech, and chestnut. These forests are usual in the western and northern mountains, and may not occur at all in drier areas.

Still higher, beginning at an elevation of three to five thousand feet, depending on the local climate, the coniferous forest zone takes over, dominated by fir and pine, interspersed with open meadows. In some places, cedar grows at this altitude. This is a subalpine zone, with higher precipitation taking the form of snow through much of the winter. Plant growth takes place in the summer, when local thunderstorms hover over the mountaintops. Above the forest, if the mountain is high enough, comes alpine tundra of dwarfed flowering plants and lichens, and finally the rocky summit peaks.

The vegetation I have been describing once supported a rich and varied assemblage of animals. Plants are the food producers of the ecosystem, and all animal life depends on them, including mankind. Just as people have changed plant communities in the Mediterranean Basin, particularly by destroying the forests, so also have they changed the animal distribution, usually by making species extinct. As an example, lions and hyenas roamed Mediterranean Europe well into historical times, until they were finally extirpated. Other species, both domestic animals and pests, have been introduced and spread by human activities.

The primeval Mediterranean fauna was related to that of the rest of Europe, with the addition of a number of endemic species, and some animals from Africa and Asia. Animals depending directly on plants for food included such large herbivorous mammals as wild sheep, goats, cattle, boars, donkeys and horses, European bison, and red, fallow, and roe deer. The elephant occurred in North Africa in early times, while ostriches ranged through Mesopotamia and Syria. Smaller plant-eaters like hares, porcupines, squirrels, and mice are joined by seed-eating birds like finches, pigeons, and sparrows, and numerous insects, from bees, beetles, butterflies, and moths to the musical cicada, beloved by the poets, and the destructive locust. Several land snails are native to the Mediterranean.

On the next level of the ecosystem are the animals that prey on other animals, obtaining their food indirectly from the producing plants through one or more additional steps in the food chain. Large predators like the lion, leopard, lynx, bear, hyena, and wolf, and the smaller fox, wild cats, the weasel tribe, and various snakes were found in the area. As of this writing, there are still jackals in Greece. Eagles, owls and other raptorial birds, vultures, ravens, and magpies are well known. Also part of this general group are the insectivores, including hedgehogs, bats, lizards and frogs, numerous birds such as swallows, thrushes, warblers, nightingales, starlings, and the crested hoopoe, and some of the insects themselves. Praying mantises, wasps, ants, and, unfortunately, mosquitoes, lice, and fleas are common. Other arthropods, such as scorpions, spiders, and centipedes, consume animal material. The Barbary ape, something of an omnivore, was common in the southern coastal areas.

We should note here that all animals and plants, before and after death, may also serve as nutriment for the decomposing molds and bacteria.

Another great section of the ecosystem is found in the waters of the Mediterranean Sea. Here, too, life depends on food producers such as the algae and phytoplanktons, and on nutrients washed down from the land. Most sea life is found in the upper layers where light penetrates. The Mediterranean reaches a depth of seventeen thousand feet, but in most places it is somewhat shallow. It is high in salt content, relatively warm, and lacks important currents to bring up nutrients from the depths and take oxygen down. While not deserted, the deeper layers of the Mediterranean are poorer in life than those of some other seas. The upper reaches contain many valuable fishes, but the most important food fishes of northern Europe are absent. Anchovies, sardines, tunnies, swordfish, flying fish, and sharks were known to the ancients. Octopus and squid, crabs, lobsters, nautiluses, and the "purple" murex, along with tritons and other shellfish, were used in food, dyes, artistic motifs, and even musical instruments. Spiny sea urchins encrusted the rocky but more often destructive. When human beings, through individual and collective actions, damaged the ecosystem, they damaged themselves and brought their civilizations into decline, because the existence and welfare of human societies depends upon maintaining a balance with nature.

important. The true affinities of human beings are with other living things. Mankind is not alone in the world.

Human beings are now generally believed to have lived in central East Africa about two or three million years ago. It is evident that early people were closely dependent upon nature in a much more direct and intimate way than later, civilized people. They had to take their daily food, drink, and shelter right from the natural environment. They competed for territory with one another and with other large predators. Sometimes they were prey, too, but their increasing ability to devise hunting tools gave them the upper hand over larger animals. While the earliest human beings may have eaten mainly small, easily gathered things like turtles, shellfish, and berries, by the time of the Ice Age people were hunting and killing the largest and most dangerous land animals.

The people who first came into the Mediterranean Basin were hunters and gatherers, armed with weapons of stone, wood, antlers, and bone. Their level of material culture is called Paleolithic, or Old Stone Age. Since they were directly dependent on wild animals and plants for food and clothing, their numbers and the size of their groups were limited. As hunters and gatherers, they were subject to the same controls as other omnivorous predators. Too many people could not crowd into the same territory, or the food supply would be depleted and they would have to move on. A natural balance was thus maintained between human numbers and the carrying capacity of the environment. People followed the migrations of the herds of animals they hunted. They used the materials which were found in abundance in the terrain they inhabited; stone, easily fashioned into axes, spearheads, and, later, arrowheads; wooden shafts, straightened and pointed, hardened by fire; and the bones and antlers of the animals they killed, which could be formed and sharpened into fishhooks, awls, needles, and instruments for working other materials.

At this stage of human cultural development, it might seem that the human impact upon the natural environment would have been minimal. Some anthropologists, however, believe that Paleolithic hunters produced or accelerated some far-reaching changes on the face of the earth.

There is the use of fire, for example. Some scholars hold that major grassland areas are primarily the result of burning by early people, who discovered that they could drive animals by deliberately setting fires and that burned-over areas regenerate in grass before the ecological succession restores them to brushland or forest. Grass provided a better source of food for the grazing animals which were their usual quarry, and periodic burning could keep land in the grassland stage indefinitely.

It is also believed that the pressure of Stone Age hunters may have caused or hastened the extinction of many of the animals of the Pleistocene period. The change in climate at the end of the Ice Age no doubt played a major role in this process of extinction, but it is possible that as the temperature of the earth rose, some of the animals might have found ecological niches further north if they had not been hunted. At the very least, some of them could have survived for a much longer time. To scan a list of species that disappeared in postglacial times is to see why people might gladly have hunted some of them to extinction as rival predators and dangerous enemies. The cave bear, cave hyena, and similar species are of this type. Larger herbivores such as the mammoth and rhinoceros were good sources of food, but dangerous and unlikely to be preserved, as the reindeer was, through semidomestication.

Hunters domesticated the dog. The ancestor of the dog was probably one of the smaller races of wolves found in the Near East, tamed for their usefulness in the hunt. Orphaned puppies, raised by human groups, transferred their inborn patterns of hunting in packs to their new masters. Reindeer must have been domesticated in a slow process of association. From following and hunting the migrating herds, people progresssed to controlling the movements of the herds and protecting them. As the glacial ice retreated, reindeer and the human groups dependent on them moved far to the north where they still live, in Lapland and Siberia.

People at the hunting and gathering stage of development left no written records, but it is possible to gather some impression of their attitudes toward the natural environment from their artifacts, carvings, and paintings, and from the attitudes expressed by tribes which survived into modern times with a similar hunting culture. There seems little doubt that these people regarded nature with a mixture of awe and a desire to manipulate it and secure its cooperation—an attitude in which elements of both magic and religion were present and not yet differentiated.

To the hunters, nature was the source of life. Since their welfare depended on success in the hunt, they developed rituals to assure this. They painted pictures of the animals they hunted, and slew them symbolically by casting spears at them or by painting in spears. All their ritual practices seem to have been conducted with nature symbolism. Sacred dances in which men wore the heads and horns of animals are depicted on cavern walls in Mediterranean France and Spain, and the evidence indicates that they used animal motifs in the ritual for initiation of youths into manhood. Modern primitive hunters are known to pray to and propitiate animal and plant spirits and to adopt their representations as identifying and protective totems. They seem also to have a general feeling that they hunt and gather from need only, and should not kill or destroy wantonly. They sense that the other creatures of the earth have spirits and consciousness, and power to help or hurt. This attitude should not be lightly dismissed as "anthropomorphism" or "the pathetic fallacy." It is a belief which has a genuine function in the delicate relationship between these people and the ecosystem within which they live. At their level of technology, they are indeed subject to the power of nature, and if they fail to find the proper balance with it, they will suffer.

The domestication of animals and plants caused such a deep and far-reaching change in mankind's relationship to nature that it is often called the "Agricultural Revolution," even though the process took hundreds or even thousands of years. It begins the period known as the Neolithic, or New Stone Age. The change made two new life styles possible for people: the pastoral life of herders and the settled life of farmers.

Instead of following the herds of grazing and browsing animals to hunt them, some people, with help from the already domesticated dog, began to protect them from predators and control their annual movements to good summer and winter ranges. This meant a pastoral, somewhat nomadic life. People continued to hunt at times, but rapidly came to give most of their attention to the herds of sheep and goats which provided them with a more dependable source of food, clothing, and shelter. The invention of weaving accompanied the domestication of these animals, whose hair and wool were the raw materials for the loom, giving the seminomadic herders clothes for warmth and tents for shelter to replace the ruder animal skins used by hunters.

Others discovered how to cultivate plants, particularly the cereal grains such as wheat and barley. In earlier times, in Palestine and elsewhere, people had used stone-edged sickles to reap the wild grain "they had not sown."[1] But the development of planting, cultivating, and harvesting provided a dependable food supply, which had never before been available. It both required a settled community to care for the crops and made possible an enlarged population in a smaller area. When people were tending their crops, their movements were restricted, at least during the growing season. They could sometimes go hunting, and they kept some of the same animals the herders did, including cattle and pigs, but their main concern and their main diet were the new grain crops.

The farmers lived in villages, their simple quarters constructed of stone, clay, and wood clustered together in a defensible spot, with secure pens in which to keep the animals at night. The first such villages arose in an area extending from the northeastern Mediterranean Basin eastward into Asia. The center of this area is the foothills at the northern margin of the Mesopotamian plain, in an area then consisting of winter grasslands alternating with open woodlands of oak and pistachio, well above the swampy tangles of the river valley. Westward, archaeologists have found remains of these villages, dated at about 8000 to 6000 B.C., above the Jordan Valley, on the hilly margins of the Anatolian plains, and in Macedonia, all in country of similar character. Settlements have also been found further east, in Iran. Recent discoveries may indicate that agriculture based on other plants and animals developed at about the same time or earlier in Southeast Asia.

It is in these foothill areas of the Near East, as one would expect, that the wild ancestors of the cultivated grain grasses were found, as well as the wild sheep, goats, pigs, and cattle which were later domesticated. These animals are all at home in ecotonal country where grassland, brushland, and forest interpenetrate. Sheep and cattle graze, but also seek the shelter of trees. Goats, proverbial eaters of almost anything, generally prefer to browse in shrubs and small trees, and pigs root in forest and brushland.

Human ability to change and control the natural environment greatly increased with the agricultural revolution. The herders became a force for the destruction of forests. Goats not only browse their favorite shrubs but can climb right up into trees to eat the foliage, and they eagerly consume the seedlings of trees, so that where they are constantly herded, forests cannot regenerate. The rooting habit of pigs also can damage forests and brushland. As soon as sheep were domesticated, the danger of overgrazing appeared, since they will eat grass, roots and all, and their sharp hooves tear up the sod. Cattle, though not quite so destructive, can also overgraze, and herders often set fires to encourage the growth of grass. Balancing the destructive effects to some extent was the return of nutrients to the soil in the form of manure, and the practice of transhumance, or the movement of the herds to different pastures in summer and winter.

Early farming involved tilling the soil, deliberate clearing of land for agriculture, and excessive use of the forests nearest the villages for firewood and building materials. With the removal of natural plant cover came erosion, so that these hilly districts in which men have practiced subsistence agriculture for ten thousand years or so are now desiccated, rocky, and almost denuded of useful plants. Such effects, though undeniable, were slow and cumulative. Neolithic people, whose tools were of carefully fashioned stone and sometimes later of copper, who lacked a true plow and efficient methods of irrigation, found their numbers limited by the fertility of the soil under their care. They used terracing, a form of soil conservation. All told, they still managed to remain in balance with their slowly changing environment.

The attitude of the Neolithic farmers and herders toward the natural environment was strongly magical and religious. In the large agricultural village of Çatal Hüyük in Turkish Anatolia, for example, about one out of every three rooms was apparently used for ritual purposes, judging from murals and special objects such as the decorated skulls of bulls found in them.[2] The Hopi Indians, whose way of life was at the stage of Neolithic agriculture until quite recently, devoted fully one-third of their waking hours to religious ceremonies, chiefly dances.

Neolithic religion centers around the encouragement of fertility. The well-known female figurines with their exaggeration of rich layers of fat and reproductive characteristics, often associated by modern scholars with the "mother goddess," exalt one aspect of the bounty of nature, while the dominant motif of the bull, an obvious embodiment of the masculine force of fertility among domesticated animals, represents another. Everywhere in the Near Eastern lands which are the birthplace of agriculture, the bull cult appeared. Murals of bulls painted in black and red decorate walls at Çatal Hüyük, and the skulls or horns of bulls, combined with modeled clay covered with geometric designs, project from the walls or floors of rooms used as shrines. Other animals were also celebrated as symbols of life and death; domestic sheep and pigs and wild leopards and vultures occur along with human figures and handprints. Art in the agricultural villages is much less realistic than art in the hunters' caves of former times. The farmers stylized the forms of animals and human beings, combining them with repetitious angular patterns in various colors, chiefly black, white, and red.

The religion of Neolithic peoples had a strongly ritualistic element; the same motifs and actions were repeated over and over in an attempt to encourage nature to give rather than to take away. The natural environment was the source of life; it was also the source of drought and flood, of predatory animals and death. The somber scenes of black vultures pecking at headless men contrast with those of cows with full udders and of the mother goddess giving birth, her arms raised as if in benediction.

The men of the agricultural villages and domestic herds felt as one with nature. They were painstakingly conscious of the ever-repeating changes of the seasons, and they watched the rising and setting of the sun, moon, and stars with great care. They set their great ceremonies to coincide with seedtime and harvest, and their religious observances formed a complex annual cycle which, they feared, could not be neglected without grave consequences. They realized that human beings have a place in the unity of nature. As they had to cultivate the earth, plant the seed, and save enough for the coming year if nature was to provide her increase, so their prayers, they earnestly believed, helped the rains to come, the grain to grow, and the ewes to bear lambs.

4

Early Civilizations and the Natural Environment

Cities, temples, palaces, and tombs of once-flourishing societies now lie in ruins throughout the Middle East. Here people first developed high civilizations, and here, in a particularly telling way, the surviving evidence shows that the course of history is not always that of upward human progress. The rise of civilizations depended upon the increasing ability of people to use and control their natural environment, and the downfall of these same civilizations was due to their failure to maintain a harmonious balance with nature. They suffered a true ecological disaster: not simply a change in climate—for people have weathered climatic changes before and prospered—but a disaster of their own making. This chapter will examine the successes and failures of the Mesopotamians and Egyptians in their relationships to the ecosystems in which they lived.

Mesopotamia, the broad plain of the Euphrates and Tigris rivers, borders the Mediterranean Basin on the east, and at one point north of Antioch in modern Turkey, the Euphrates comes within 90 miles of the Mediterranean coast. But the climate of Mesopotamia is very dry, with an annual rainfall of only six to eight inches. Summer temperatures of 120 degrees in the shade are not uncommon, and 137 degrees has been recorded. Most life in Mesopotamia is dependent on the water brought down from the mountains by the two rivers. The twin rivers begin in the snows of the Armenian mountains, which reach an elevation of almost seventeen thousand feet at Mount Ararat, and flow down into the head of the Persian Gulf, a distance of 1150 miles for the Tigris and 1780 miles for the more circuitous Euphrates. A major tributary stream, the Karun, flows down from the Persian mountains. The plain is an alluvial deposit of sand and silt brought down by the rivers in past geological times, and the process is still going on. This undoubted fact once caused historians to believe that at the time of the earliest civilizations in Mesopotamia, the Persian Gulf extended much further northwest, providing Sumerian cities with a seacoast, and many maps in ancient history textbooks show it that way. But careful studies by geologists in the area have shown that in spite of the sediment still being brought down by the rivers, the land area in the delta region has not increased, because lower Mesopotamia has been undergoing periods of geological subsidence.[1] The coastline today, with minor variations, is in about the same position as it was 5,000 years ago.

The world's first cities, which arose in Mesopotamia and nearby, were made possible by a changed relationship between human beings and the environment, based on a new agriculture using two important inventions: systematic irrigation and the plow. The fertile, sandy, easily turned Mesopotamian soil made the plow useful. The rivers provided the essential water, but with a flow so undependable that control by major irrigation works was demanded. The new agriculture enabled a much larger human population to live in a given area, and an increasing portion of the population, freed from the need to work the soil, could take up specialized occupations.

The earliest cities seem to have shared some of the problems which have become so annoying in their modern counterparts. Babylon, in its day the largest city of the area, had a city wall ten miles long, and even including its suburbs was consequently only of moderate size by modern standards. The evidence of narrow streets and small rooms in houses huddled within the compass of defensible walls tells us that crowding in ancient cities was extreme. Garbage accumulated in the houses, where the dirt floors were continually being raised by the debris, and human wastes were rarely carried further than the nearest street. The water supply, from wells, rivers, and canals, was likely to be polluted. Life expectancy was short, due in part to the high infant mortality. Flies, rodents, and cockroaches were constant pests. Even air pollution was not absent. In addition to dust and offensive odors, the atmosphere filled with smoke on calm days. Even today, in large preindustrial cities such as Calcutta the smoke of thousands of individual cooking fires, in addition to other human activities, produces a definite pall of smoke and dust which seldom dissipates for long. Under these unhealthy conditions, the death rate must have been high in Mesopotamian cities.

As an alluvial land, the Mesopotamian plain had no stone or deposits of metallic ore, and these had to be brought in from the mountains or imported from other countries. While this encouraged the early development of trade, it also meant that the inhabitants had to use the native materials, swamp reeds and clay, for ordinary construction. They built mighty works of baked and unbaked clay bricks—temples, shrines raised on lofty ziggurats, palaces, and walled cities—but the system of canals and other irrigation works is their most remarkable achievement. Incidentally, it is interesting to note that petroleum, the most important natural resource of modern Iraq, did not escape their notice. They used the oil which oozed forth in some places as fuel for their lamps and bitumen for waterproofing their boats.

The attitude of the peoples of Mesopotamia toward nature, from early Sumerian writings down through the Akkadian and

Assyrian literatures, is marked by a strong feeling of battle. Nature herself was represented in Mesopotamian mythology as monstrous chaos, and it was only by the constant labors of people and their patron gods that chaos could be overcome and order established. Mesopotamian gods, though they retained their earlier character as nature deities to some extent, were primarily figures which sanctioned order, guarded the cities, upheld government and society, and encouraged the construction of works which would reproduce on earth the regularity of heaven. The order of heaven was quite apparent to the Mesopotamians, who developed both astrology and astronomy to a high degree and noticed that the motions of the moon and sun, stars and planets are constant and predictable. The labors of the Mesopotamian hero-god Enlil, or Marduk, in slaying the primeval monster of chaos, Tiamat, and creating the world out of her sundered body, reflected the labors of the Mesopotamians themselves, who built islands in the swamps, raised their cities above the flood plains, and irrigated desert stretches with an orderly series of well-maintained canals. They planned their cities so that the major streets of Babylon, for example, crossed each other at right angles in a regular grid pattern, and they laid out their canals in the same way wherever possible. Left to itself, Mesopotamia would have remained a land at the mercy of the capricious river and the merciless sun, in a precarious, shifting balance between tangled marsh and parched desert. But careful works of irrigation conquered sections of that land and won rich sustenance from its basic fertility. Thus a Mesopotamian king could list the construction of a new canal, along with the defeat of his enemies in battle, as the major events of a year of his reign.

Mesopotamians had a well-developed sense of the distinction between the tame and the wild, between civilization and wilderness. The proper effort of mankind toward wild things, they believed, is to domesticate them. They did this with such native animals as the donkey and the water buffalo, in addition to keeping the cows, pigs, sheep, and goats already known to their ancestors. They learned the uses of the palm tree and planted it widely. Animals which could not be truly domesticated were hunted—some, like the lion, to extinction. In the Epic of Gilgamesh, Enkidu was presented as a man of the wild, a friend and protector of beasts. But when he had been tamed by womanly wiles, his former animal friends feared and fled from him. One of the great feats of Gilgamesh and his now-tamed companion was the slaying of Humbaba, the wild protector of the cedar forests in the west, and the seal set upon the defeat of Humbaba was the subjugation of the wilderness; the trees were felled for human use. This ancient legend described an actual ecological event; the cedars of Lebanon, after centuries of exploitation and export to all the surrounding lands, were completely destroyed except for a few small groves, leaving their mountain slopes open to severe erosion.

The Mesopotamians also displayed curiosity toward and interest in the natural environment. They compiled many lists, which survive on clay tablets, of animals, plants, and minerals. These might be regarded as a step toward genuine scientific classification. However, the lists always classify natural things on the basis of the uses to which they were put by man. Mesopotamian thought was intensely practical and anthropocentric. This principle was illustrated in a Sumerian legend which told how the god Ninurta fought a war in which some stones helped him and others opposed him. Ninurta rewarded the former by making them jewels and semiprecious stones, while the latter he punished by making them into paving stones and thresholds, trodden under foot by people.

The Mesopotamian fondness for building high ziggurats and towers has been seen as a compensation for the flatness of the land by people who had moved in from hilly regions and longed for the forests and mountains of their former homes. This

irrigation and cultivation until they in turn suffered the same salinization. A similar process occurs in many places where deserts have been irrigated, as in the Imperial Valley of California today, where the best efforts of modern technology have barely been able to combat it. Salt would not have accumulated in well-drained soil, but in Mesopotamia the problem of drainage was especially difficult. Silt and mud carried by the rivers and canals settled out rapidly, so that constant dredging was necessary to keep the canals flowing. Excavated mud piled up along the sides of the canals to a height of thirty feet or more, serving as a barrier to drainage. Eventually the river level was raised well above the surrounding country. The natural remedy to this, flooding and a major shift in the course of the rivers, was catastrophic whenever it happened.

It is significant that the first urban societies were also the first societies to abandon a religious attitude of oneness with nature and to adopt one of separation. The dominant myth and reality in Mesopotamia was the conquest of chaotic nature by divine-human order. Such societies, it must be noted, were ultimately unsuccessful in maintaining a balance with their natural environment.

Far to the east of Mesopotamia, but in contact with it, another civilization flourished in the Indus River Valley. It is mentioned here because many scholars say that it fell because of mistreatment of its fragile semidesert environment. While Mesopotamian cities were built largely of sun-dried clay bricks, the Indus Valley cities used baked bricks almost exclusively, and great quantities of wood were required to fire them. This, combined with other uses of wood, produced widespread deforestation, while grazing of cattle, goats, and sheep further reduced the vegetative cover. The results included desiccation, flooding, and erosion. Some authors also theorize that the dust blown from the dried, denuded land produced a permanent layer of dusty haze in the atmosphere over the Indus Valley, which actually altered the climate by causing a temperature gradient that shifted the monsoon rains to the east, out of the area, and also caused premature seeding of the clouds which did develop in the area, further reducing the rainfall. Other scholars postulate a series of disastrous floods, which could also have been caused by deforestation.

The attitude of the Indus Valley people toward nature is virtually unknown. They did represent animals on their stamp seals, but their language has not yet been deciphered. Whatever their attitude, their practices may have made them a prime early example of the principle that human societies which fail to live in harmony with the natural environment eventually disappear or change beyond recognition.

Without the Nile River, Egypt would be part of the Sahara Desert. The facts of Egypt's climate speak for themselves. The average annual rainfall at Cairo is one inch, and southward in Upper Egypt there may be one shower every two or three years. The average temperatures are typical of the desert belt; in the valley they vary from about 50 degrees to 110 degrees on the average, while on the adjoining desert the range is even more extreme, varying from below freezing to above 120 degrees.

The Nile is the world's longest river. Rising in the mountains of Ethiopia and the lakes of East Africa, it flows northward through the swamps of the southern Sudan and then begins its course as an exotic river in the desert. Its total length is thirty-five hundred miles. Spring and summer rains on the headwaters of the Blue Nile and its tributaries, which always occurred at the same time of the year, reaching their height in September, fed the annual floods of the Nile, which brought both moisture and new alluvial soil to the fields of Egypt and were one of the major environmental influences in that country.

During flood stage, the Nile usually carred fifty times as much water as at its lowest stage.

The contrast between the land watered by the Nile and the desert which borders it is abrupt and extreme. The ancient Egyptians recognized this, dividing the earth into two parts, the fertile black land of Egypt and the dry red land of the hostile desert. The black land, in turn, had two major divisions, the long, narrow strip of Upper Egypt, from one to thirteen miles wide, often guarded by high cliffs, and the broad delta of Lower Egypt, containing two-thirds of the arable land of the country.

In very ancient times, much of the black land was covered by marshes which supported an amazing variety of plant and animal life, including millions of water birds in the Biblical "land of whirring wings."[2] The Egyptian crocodile and hippopotamus are only the best known of the vast assemblage of wild animals. The marshes also provided the papyrus reed, subject to many uses, including the manufacture of paper.

The agricultural civilization of Egypt depended on harnessing the Nile's annual flood and distributing its waters and its fertile load of silt through the fields by a series of canals and basins with the use of the shadoof, a simple water-lifting machine consisting of a bucket with an arm and counterweight attached. The Nile's flood was regular and predictable, although its height varied from year to year, a very low Nile producing drought and famine and a very high Nile devastating the works of man and even eroding the soil. Still, most years brought a moderate, useful flood.

Egyptian attitudes toward nature reflect the dependable periodicity of their natural environment. Their gods were deities of nature, intimately sharing the characteristics of the animals and plants which were their attributes. Ra, the sun god, was worshiped beyond the others. His movements were regular, and all nature responded to them. When he rose, day came and life flourished, and when he set, the Egyptians associated the failing light with death, but a new dawn inevitably followed. The Egyptians calculated the solar year of 365 days, and noticed that the Nile's flood depended on the sun's cycle. Another major god, Osiris, represented the dying and rising vegetation, intimately associated with the Nile and the sun. All Egyptian gods represented aspects of the natural world, and most of them were conceived as friendly to mankind.

In the temples and sacred precincts, the Egyptians protected animals and plants which embodied the presence of the gods, and often gave them divine honors. For example, crocodiles sacred to the god Sebak were kept in the lake at Shedet, fed with offerings, and even decorated with jewelry. Other animals were generally venerated throughout Egypt, such as the cat of Bubastis and the ibis of Thoth.

To the Egyptians, all nature was animate and filled with gods. In the paintings, sculpture, and objects of daily life made by them, an artistic joy in nature can be sensed. Their hieroglyphics include many animals, birds, and plants, often represented with great attention to natural detail. Columns bore capitals representing the lotus and papyrus plants. Egyptian poems enumerate and glorify the appearance and workings of the natural world:

> All beasts are content with their pasturage;
> Trees and plants are flourishing.
> The birds which fly from their nests,
> Their wings are stretched out in praise to thy
> spirit.[3]

This poem comes from the time of Ikhnaton the monotheist, when art and literature emphasized the natural, but the Amarna period (named after Ikhnaton's residence city) in this respect simply reasserted a tendency which was present in Egypt from the earliest surviving evidence.

Thus it is not surprising to discover that well-to-do Egyptians loved gardens, that they planned them carefully with symmetrical beds of flowers and shallow pools of water, and that they collected vegetables, herbs, vines, and fruit and shade trees to plant in them.

Practical knowledge of the workings of those parts of the environment directly useful to them the Egyptians had in abundance, and with it they maintained a flourishing agriculture throughout their history, except for the worst periods of invasion and internal unrest. Then the canals might go unrepaired and the desert and swamp readvance at the expense of the cultivated fields.

Egyptian science is known from treatises on mathematics, astronomy, and medicine. The medical writings include some directions regarding the use of plants and drugs in treating ailments.

The history of the Egyptian environment in antiquity is marked by a great reduction in the numbers and abundance of wildlife. This was primarily due to the conversion of marshes into fields, but partially also due to hunting. Egyptians from the pharaoh on down hunted water birds and animals in the remaining wetlands, and pursued lions, wild cattle, deer, and antelope in the nearby desert. Today even once-abundant species are seldom seen.

As Egypt was never forested, most wood had to be imported from Lebanon or the Upper Nile. The chief material for major construction, stone, was abundant in Egypt and widely quarried. Copper and tin were mined in Sinai and other desert margins or imported from abroad.

The regularity of the Nile saved Egypt from some of the problems of Mesopotamia. The floods provided annual drainage, and salinization was not widespread in Egypt. In fact, Egypt continued to produce food surpluses throughout the ancient period and was a major exporter of grain to Greece and Rome.

All told, the unique environment of Egypt tended to shelter it from some of the bad effects of ecological change felt elsewhere in the ancient world, and helped to ensure its long continuity as a relatively conservative civilization.

The Persians had a unique view of the natural world which gives them an important place in any consideration of ecology in ancient times. Their empire eventually stretched from India to Egypt and Greece, altering and influencing much of the Mediterranean Basin. The eastern neighbors of Mesopotamia, the Persians inhabited an arid, mountainous region where agriculture required extensive irrigation. For this purpose, the Persians developed underground channels, called *qanats,* through which water could flow from the aquifers without too much loss by evaporation. Cities had developed in this area before the arrival of the Persians themselves, and besides settled irrigation agriculture, many of the inhabitants practiced seminomadic herding of cattle.

The Persian religion, based upon an Indo-European pantheon of nature deities, developed a strong sense of reverence for the elements of the natural world. Earth, water, and especially fire were regarded as sacred in themselves, and much of Persian religion was concerned with the need to keep them free from ritual pollution. The Persians had many rules concerning cleanliness, and those who broke them could be severely punished. Water was to be kept in a state of pristine purity, whether it was flowing or standing in a lake or well. Sewage of any sort, whether urine or excrement, or even hair or fingernail parings, was not permitted to enter water, although water might itself be used as an agent of purification. The worst sort of pollution in Persian eyes resulted from contact with dead bodies of human beings or animals, and anyone seeing one of these in the water was duty bound to remove it and perform the necessary purifications. The earth was also sacred, and burial of dead bodies was considered the worst possible violation of the will of the gods. Since fire, an object of great veneration, could not be polluted either, bodies were simply exposed in rocky places or in special towers to be eaten by "unclean" animals such as wolves and vultures.

Along with the elements, the Persians worshiped sacred plants, animals, and stars. Certain human activities, such as agriculture, were regarded as acts of reverence to the earth which made the earth happy and fruitful. Domestic animals were given reverent care.

When the prophet Zoroaster reformed the Persian religion, he left intact the basic reverence for nature and the ritual maintenance of purity associated with it. He emphasized the dualism of Persian thought, however, including a certain ambivalence toward nature. While the elements are pure and must remain uncorrupted, the creatures of the earth are divided into two classes, good and evil. Ranged on the side of Ahura Mazda, the god of light and goodness, according to Zoroastrian thought, were all the good creatures, such as dogs, cattle, trees, and the sun itself, whose very rays helped to purify. On the other side, with Ahriman, the evil prince of darkness, were noxious creatures like wolves, snakes, demons of disease, and flies. Killing such creatures was regarded as an act of merit, so that condemned criminals were sometimes given the task of killing a certain number of them as a means of expiating their guilt.

In the Persian view, then, people were regarded as responsible for their actions in regard to the natural environment. They were seen as coworkers of the good Creator, charged with maintaining the purity and fruitfulness of the earth in defiance of the attempts of the forces of evil to pollute and destroy the earth and its good creatures.

It could be expected that people who observed the Persian religious rules would be healthier and cleaner than their neighbors in the ancient world, and this may have been the case.

It would be nice to add that the Persian attitude of human responsibility toward the natural environment retarded the deterioration of the Persian landscape, but this does not seem to be true. The hillsides of Persia, like those of Lebanon, were deforested and subjected to erosion. Persian fields, like those of Mesopotamia, suffered salinization. Wildlife was gradually eradicated. Some of these results may be put down to the attitude of warfare toward the "evil" part of nature, and the belief that agriculture is invariably good. But more than this, the Persians illustrate a general principle of human ecology, that is, that a good attitude toward nature is not enough. Combined with a good attitude must be accurate knowledge of the workings of nature and the ability to control and direct human impact upon nature in channels which will help, rather than hinder, the balance of nature. The Persians' considerable experience with the natural environment is reflected in many of their wise rules. But they had no science worthy of the name, and their level of technology and social control, while high enough to be one of the wonders of the ancient world, was insufficient to put into effect the limited ecological insights which were contained in their religion.

5

Ancient Israel and the Natural Environment

The importance of Israel in the history of the relationship of mankind to the environment results from the wide influence of the Judaic religion and its implicit attitudes toward the natural world. Israel was a very small country, and for much of its history was at the mercy of the great Near Eastern empires, so that the evident deterioration of the Holy Land is not the result solely of the attitudes and practices of the Jews. But the Judaic system of thought has made its impact felt both directly and also through Christianity and Islam upon a major portion of the human race, and consequently has affected mankind's treatment of the earth.

The major distinctive contribution of Israel to mankind is, of course, the idea of ethical monotheism. The God of the Jews was one God only, and tolerated no other gods beside him. He was not regarded primarily as a nature deity. True enough, God was the creator of all the earth and all that is in it, and the sky and stars as well. His creative power could be seen in the things that he had made, and his actions of providence and miracle could be seen in the world of nature. But God was above and beyond his creation, ruling it from on high. He might ride upon the storm, but he was not the storm. He had formed the earth, and caused the waters to flow, but he was not the earth or waters. He was not in the earthquake, wind, and fire, though he might send them. Thus the transcendent God of Israel could be set over against his creation in a way that was impossible with the many nature gods of surrounding peoples.

The Jews loved nature, and saw great beauty in it. The poetry of much of the Song of Songs reflects this, as in this lyrical passage describing the coming of spring:

> The winter is past, the rain is over and gone.
> The flowers appear in the earth, the time of singing has come,
> And the voice of the turtledove is heard in our land.
> The fig tree puts forth its figs, and the vines are in blossom;
> they give forth fragrance.[1]

Other passages—many of the psalms, like the great 104th; the story of creation in Genesis; and the words of the voice from the whirlwind in Job—are some of the finest descriptions of nature in all ancient literature.

When the Jews looked at nature, they saw it not only as beautiful in itself, but as a manifestation of the power and majesty of the Creator. "The heavens are telling the glory of God; and the firmament proclaims his handiwork,"[2] sang the psalmist. The sea, the trees, the mountains, and all living creatures give glory to God. "The whole earth is full of his glory."[3] The magnificent 148th Psalm calls the roll of the creatures and exhorts them to praise God. The natural environment, while not regarded as identical with God or an extension of his being, could serve as evidence of his mighty presence, and was therefore full of significance.

When God created the world, he repeatedly called it good. "And God saw everything that he had made, and behold, it was very good."[4] Unlike the Persians, the Jews conceived no evil counter-creator, and thus everything in nature was regarded as good in itself. True enough, some animals were regarded as ritually unclean; their flesh was not eaten, and contact with them was avoided, but in their proper place in the world's order, they were among God's good creatures. God knows all his creatures, the Jews believed, delights in them, and gives them their food in due season.

"The earth is the Lord's," the Biblical writers repeatedly asserted, "the world and those who dwell therein."[5] The Judaic teaching was that God is the ruler of the universe and retains the ultimate dominion over the natural environment. "In his hand are the depths of the earth; the heights of the mountains are his also. The sea is his, for he made it; for his hands formed the dry land."[6] He not only created the world, but afterward maintained his sovereignty over it.

Mankind's place on earth, according to the Judaic view, is subordinate to God's ultimate dominion. God created human beings, male and female, in his own image, and gave them dominion over "the fish of the sea and over the birds of the air and over every living thing that moves upon the earth."[7] but only as God's deputies. Human beings are not the lords of creation, free to do with the earth as they please, but God's viceroys, responsible to God for their actions. The Judaic emphasis upon ethical conduct in obedience to God's law applied also to human treatment of the natural environment. The grant of dominion was not a license to kill, exploit heedlessly, and pollute, and was not understood as such by the ancient Jews, although later Western thought did indeed take it in that distorted sense.

Compared with the nature cults of the surrounding peoples, Judaic thought regarded the natural environment with greater respect and care. True enough, their neighbors identified their gods with natural forces, but their attitude toward them was manipulative, and worship was intended to encourage good returns to the worshiper. To the Jews, the earth and its creatures were a trust from God, to be protected and used wisely.

The Book of Genesis says that "the Lord God took man and

put him in the garden of Eden to till it and keep it."[8] Mankind was given a task by God, to tend the unspoiled natural world by the careful practice of agriculture. Even then, certain trees were to be preserved from human use. When mankind failed that trust, creation fell too, but the task was left the same. Mankind was still to till and keep the earth, but now in toil, sweat, and pain.

One of the punishments meted out to unfaithful mankind after the Fall was that the earth would bring forth unpleasant weeds, thorns, and thistles. In many places throughout the Hebrew Bible, the aspect of the natural world serves as a moral commentary upon the conduct of mankind. When people do evil, when they fail to obey the will of God, render unjust judgments, and mistreat those who are helpless, "the ground mourns," "the grain is destroyed," and "the vine . . . and all of the trees of the field are withered."[9] But when God is pleased with his people, when they have suffered enough and manifested their faithfulness to him, nature expresses God's goodness. "The mountains shall drip sweet wine, and the hills shall flow with milk."[10] "The desert shall rejoice and blossom."[11] These are the results, however, not of wise husbandry and sound agricultural practices or the lack of them, but of social and political justice or its absence.

The ancient Jews regarded pastoral activities as the good, normal occupation of mankind. Their early history is marked by a sense of conflict between this way of life and settled agriculture; in Genesis, Cain's offering of the fruit of the ground was rejected by God, but Abel's gift of lambs from the flock was accepted. The hostility between urban and rural cultures is also not unknown in the Bible. Yet work on the land was not considered degrading; it could be said of a king that "he loved the soil."[12] As a good work, Abraham planted a tree by a well. The Jews built dams, barriers, channels, and storage cisterns for the conservation of water, particularly in the drier parts of

Palestine, depending as they did upon the "former and latter rains"[13] of the late fall and early spring. Lacking great rivers, they found "their Nile in the sky,"[14] as the Egyptians noted with some astonishment.

The Hebrew Bible contains many commandments concerning the proper use of the land. Prominent among them are those dealing with the Sabbath. Not only was one day in seven set aside for people and domestic animals to rest, but one year in seven was made a fallow year for the land, the produce it brought forth of itself being reserved for the poor. Prohibitions against plowing with an ox and donkey together and against sowing a field with more than one type of seed at a time are related to the desire for ritual purity which is as strong among the Jews as among the Persians. Prohibitions against uncleanness would also have applied to contamination of springs, streams, and the land. Like the Persians, the Jews abhorred blood and dead bodies more than the commoner forms of pollution, but provisions were made for the careful disposal of the latter as well. Other provisions of the Jewish law seem, however, to be motivated by kindness and a respect for other creatures. For example, it was forbidden to muzzle an ox while he was treading the grain, or to take a mother bird which was sitting on eggs or the young.

The Jews seem to have possessed considerable practical ecological insight, and they might have prevented some of the extreme damage which has been sustained by the Holy Land in deforestation, erosion, and exhaustion of the soil. Because their homeland was periodically disrupted by military campaigns and exploited by a series of conquerors, however, it is impossible to say what the Jews by themselves might have done for their land in antiquity.

6

Greek Religion and the Natural Environment

The attitudes of the early Greeks toward nature were shaped by their religion. They saw the natural environment as the sphere of activity of the gods. Greek religion was in large part the worship of nature, and the old Greek gods were essentially nature deities. The gods ruled nature, they appeared in it, they acted through it; therefore, human activities which affect the environment often were seen as involving the interest and reaction of the gods.

Zeus was a weather god, wielding the dreaded thunderbolt. He made the wind blow favorably or disastrously, moved the clouds, stirred up storms, thundered, lightened, and roused the waves of the sea. Of a particularly foul spell, Homer said, "Zeus rained the whole night through."[1] Poseidon, originally the underworld god of springs, the earthquake god who shook the mountains, became the preeminent god of the sea. Athena's patronage of wisdom and defensive warfare were importations; her earlier concerns were the owl, the serpent, and the sacred olive tree, all objects depicted in association with a female deity on ancient Minoan seals. Artemis, far from being only the divine huntress, was from of old "Artemis of the wild wood," the protectress of wild animals and the guardian of the wilderness, bearing the very ancient title Potnia Theron, "Mistress of

Beasts."[2] She was a fertility goddess who encouraged the multiplication of birds and animals. But the true patroness of the earth's fertility was Demeter, the goddess of plants. The growing grain, the Greeks believed, was Demeter at work.

The gods and goddesses made their homes in the wilderness, and when they invaded the world of human beings, it was out of the wilderness that they seemed to come. They loved the mountains; Mount Olympus, their chief dwelling, was the highest peak and one of the wildest in Greece. When they were not on Olympus, Zeus and Hera made "many-fountained Mount Ida, the mother of wild beasts,"[3] their bower. Apollo and the Muses frequented the forests, crags, and springs of mountains like Helicon and Parnassus.

Greek religion had a strong sense of natural locality. The presence of the gods was felt in places of natural attractiveness and significance. The great gods and lesser deities haunted particularly wild and beautiful locations such as springs, caves, groves of trees, and places with wide and inspiring views. Specific features of the landscape, or their sprites, figured as lesser gods and goddesses in their own right. Rivers ranked among the immortals, as did the winds, and the Greeks sometimes prayed to them.

Every one of the great national shrines and temples of the Greeks had a location and an orientation which was dictated by its natural setting. Those who have been to Greece know that the oracle of Delphi was located in a spot which commands one of the most spectacular natural scenes on earth, looking up at the "shining cliffs" of Parnassus and down the canyon of the Pleistos River as far as the flashing sea. The healing shrine of Asclepius and its giant theater are set in a natural amphitheater at Epidaurus. The mountain view from the ancient shrine of Zeus at Dodona is majestic. Without any important exception, the great oracles where the gods spoke to people, the temples, the theaters where divine pageants were enacted, and the stadiums should be restored to her mother if she had not eaten anything in the underworld. It was discovered that Persephone had eaten only four tiny pomegranate seeds, and so a compromise was reached. Persephone would spend four months underground each year, during which the crops would not grow, but during the other eight months, she would live with her mother, who would then cause the seeds to sprout and the world to be clothed once again in living green.

The myth was understood by the Greeks to signify the origin of the seasons. The "four months underground" are the dry season when practically nothing grows in Greece. But the Mysteries of Eleusis went beyond this simple symbolism to identify the life and death of men and women with the dying and rising of the natural world and its goddesses which can be seen every year. People die and are buried in the earth, as seeds are planted in the soil. But as seeds send forth shoots in response to healing moisture, they believed, those who were initiated into the mysteries would flourish again and live a happy, immortal life in companionship with the gods in the next world. The mystery of Eleusis was the identification of mankind with the natural world of growing plants, of sowing and harvest, in the great, never-ending cycle of being.

At Eleusis and in their other great festivals, the Greeks celebrated the cycle of the seasons. They felt a strong sense of an underlying order and balance in nature. They believed that the immortals, or a principle of justice even higher than the gods, operated to keep everything in its correct place, spatially and temporally. "The immortals have appointed a proper time for each thing upon the earth, the giver of grain."[6] The Greeks hated the ideas of limitlessness and chaos. In the beginning, says Homer, the world was divided into three parts, and each was given by lot to one of the three brother gods to rule; the sea to Poseidon, the underworld to Hades, and the atmosphere to Zeus, but the earth and Mount Olympus were to be held in common.

Beasts."[2] She was a fertility goddess who encouraged the multiplication of birds and animals. But the true patroness of the earth's fertility was Demeter, the goddess of plants. The growing grain, the Greeks believed, was Demeter at work.

The gods and goddesses made their homes in the wilderness, and when they invaded the world of human beings, it was out of the wilderness that they seemed to come. They loved the mountains; Mount Olympus, their chief dwelling, was the highest peak and one of the wildest in Greece. When they were not on Olympus, Zeus and Hera made "many-fountained Mount Ida, the mother of wild beasts,"[3] their bower. Apollo and the Muses frequented the forests, crags, and springs of mountains like Helicon and Parnassus.

Greek religion had a strong sense of natural locality. The presence of the gods was felt in places of natural attractiveness and significance. The great gods and lesser deities haunted particularly wild and beautiful locations such as springs, caves, groves of trees, and places with wide and inspiring views. Specific features of the landscape, or their sprites, figured as lesser gods and goddesses in their own right. Rivers ranked among the immortals, as did the winds, and the Greeks sometimes prayed to them.

Every one of the great national shrines and temples of the Greeks had a location and an orientation which was dictated by its natural setting. Those who have been to Greece know that the oracle of Delphi was located in a spot which commands one of the most spectacular natural scenes on earth, looking up at the "shining cliffs" of Parnassus and down the canyon of the Pleistos River as far as the flashing sea. The healing shrine of Asclepius and its giant theater are set in a natural amphitheater at Epidaurus. The mountain view from the ancient shrine of Zeus at Dodona is majestic. Without any important exception, the great oracles where the gods spoke to people, the temples, the theaters where divine pageants were enacted, and the stadiums whose athletic contests were devoted to the gods were located where nature had first produced a fine wild scene. The Greeks were sensitive to the natural environment as it existed unchanged by human activities, and they responded to the beauty of places like these by dedicating them to the worship of the gods whose presence they felt there.

The association between the gods and trees was particularly close in the Greek mind, and all Greek altars and places of worship were originally outdoors in groves of trees. The practice of setting aside such a grove, called an *alsos,* as a sacred area of land, or temenos, of a deity, to be used only for worship, was widespread. Acts of worship generally took place outdoors; temples were only shelters for the images of the gods and votive offerings, with no place indoors for sacrifice. Carved Minoan seals show goddesses seated under the branches of trees. There was a palm tree next to Apollo's altar on Delos, and a laurel at Delphi. An olive tree stood next to Athena's shrine on the Acropolis of Athens, where to protect the tree no goat was allowed to come except once a year, for sacrifice. All oaks were sacred to Zeus, especially at Dodona, where his voice could be heard in the rustling leaves. Probably the Greeks first worshiped in groves and only later built temples within them; it is certain that the association of the two was so deep that temples were always set in groves of trees, and when a grove was not available, as was the case on the rocky summit of the Acropolis when the Parthenon was built, holes were excavated in the solid rock and trees were planted in rows to flank the temple. The Greeks could not conceive of a sacred area without trees. The result of this attitude was the preservation of the sacred groves from fire, grazing, and the ax, so that they remained like parks in a seminatural state, and the trees in them grew huge with age. The cypress trees at Psophis, called the "Maidens," reached such a height that they overshadowed a nearby hill.[4] When the fertile plain of Crisa, near Delphi, was dedicated to the god Apollo, it was allowed to revert to a wild state, and the attempt of the Phocians to cultivate it was the official cause of the Third Sacred War. In later times the countryside was dotted with large groves of trees saved from the general denuding of the landscape, even long into the Roman period, when they aroused the wonder of travelers.

Not only trees, but also wildlife, were protected in this way. Hunting in sacred groves was forbidden. At Pharae, fish sacred to Hermes could not be caught.[5] Athena's fortress in Athens sheltered small owls and a half-legendary snake. The *Odyssey* recounts the famous story of the wild cattle sacred to Helios, the sun god, whose slaughter brought swift punishment on the guilty. The gods were believed to appear in the forms of birds and animals at times, and to send birds and other wild creatures as omens, warnings to those who could interpret them correctly, so it is not surprising that creatures living in the sanctuaries were protected. From religious motives, then, the Greeks set aside areas which, though very numerous, were usually small and in later Greek history lost most of the wilderness character they originally possessed. It is probably stretching the point to see these as national parks in anything like the modern sense, but at least it is clear that the Greeks preserved certain natural areas.

Greek religion recognized mankind's oneness with nature. This attitude was exemplified in the initiation ceremony at Eleusis, where thousands of Greeks saw and heard the enactment of the mysteries of the goddess Demeter. Demeter, the guardian of growing crops, was identified with grain, the staple life-supporting food of the Greeks. An ancient myth said that Demeter's daughter, Persephone, was once seized by Hades and carried off to his kingdom in the underworld. Demeter, in her agonized search for her beloved child, caused all plants to stop growing, threatening the destruction of life on earth. The prayers of people then swayed Zeus to decree that Persephone should be restored to her mother if she had not eaten anything in the underworld. It was discovered that Persephone had eaten only four tiny pomegranate seeds, and so a compromise was reached. Persephone would spend four months underground each year, during which the crops would not grow, but during the other eight months, she would live with her mother, who would then cause the seeds to sprout and the world to be clothed once again in living green.

The myth was understood by the Greeks to signify the origin of the seasons. The "four months underground" are the dry season when practically nothing grows in Greece. But the Mysteries of Eleusis went beyond this simple symbolism to identify the life and death of men and women with the dying and rising of the natural world and its goddesses which can be seen every year. People die and are buried in the earth, as seeds are planted in the soil. But as seeds send forth shoots in response to healing moisture, they believed, those who were initiated into the mysteries would flourish again and live a happy, immortal life in companionship with the gods in the next world. The mystery of Eleusis was the identification of mankind with the natural world of growing plants, of sowing and harvest, in the great, never-ending cycle of being.

At Eleusis and in their other great festivals, the Greeks celebrated the cycle of the seasons. They felt a strong sense of an underlying order and balance in nature. They believed that the immortals, or a principle of justice even higher than the gods, operated to keep everything in its correct place, spatially and temporally. "The immortals have appointed a proper time for each thing upon the earth, the giver of grain."[6] The Greeks hated the ideas of limitlessness and chaos. In the beginning, says Homer, the world was divided into three parts, and each was given by lot to one of the three brother gods to rule; the sea to Poseidon, the underworld to Hades, and the atmosphere to Zeus, but the earth and Mount Olympus were to be held in common.

The gods rule the natural environment and manipulate it for purposes of their own, which powerfully affect the fortunes of mankind, for good or for ill.

When everything holds its proper time and place, then all is right on earth and in heaven, and justice reigns. To overstep the bounds, to attempt to change the natural order of things, is to do injustice, to upset the world and the gods. Human beings, the Greeks thought, tend to violate the order of the universe whenever, in their pride, they try to make major alteration in what is already present in the natural environment. Canals across isthmuses, for example, were strongly discouraged because they would have made islands of what were naturally peninsulas. The wall, ditch, and palisade that the Greeks had built to face the walls of Troy, according to Homer, provoked the gods to anger, and Poseidon later destroyed every trace of the structure. When the Persian king Xerxes invaded Greece, the Greek historian Herodotus recorded as evidence of the pride that goes before a fall the facts that he had a bridge of boats built across the sea at the Hellespont and ordered the waves whipped when they broke it, that he caused a canal to be cut through the peninsula of Athos, and that his army drank rivers dry and set forests on fire. When the Colossus of Rhodes fell in an earthquake, it was taken as evidence of divine anger at human presumption, and a direct order of the oracle at Delphi forbade its reerection.

The Greeks' belief that the gods, in upholding the order of nature, would punish the transgressor supported an impressive list of taboos against pollution of various kinds. Hesiod, who as a farmer had a certain hard-bitten reverence for the earth, recorded strong prohibitions against the contamination of rivers and springs by human wastes. It might be argued that rules like these were not the result of rational recognition of hazards to health and the environment, but were dictated in an arbitrary way by purely ritual considerations. Homer, it is true, regards

Achilles' pollution of the River Scamander with blood and corpses as an offense against the river god himself, and Scamander's wrath takes an appropriate form. Like any choked river, he raises a flood and threatens to drown Achilles and bury him in sand, gravel, silt, and slime. At another time, the god Apollo sees Achilles dragging the mutilated, bleeding body of Hector behind his chariot and remarks, "Lo, in his fury he is dishonoring the silent earth."[7] But such religious taboos are more often than not the expression of deep insights into the nature of things. In any case, their effect in Greece seems to have been in the direction of preserving rather than destroying the environment. For the ancient Greeks, the natural environment was endowed with living, divine presences which maintained its order and resisted the more ill-considered actions of human beings.

But the order of nature can be broken in many ways. To the Greeks, religion was not an isolated aspect of life; it included and bore upon everything else. The gods, offended by human injustice on what would now be called the social level, could manifest their wrath in natural disasters.

> And even as beneath a tempest the whole black earth is oppressed, on a day in harvest-time, when Zeus poureth forth rain most violently, whenso in anger he waxeth wroth against men that by violence give crooked judgments in the place of gathering, and drive justice out, recking not of the vengeance of the gods; and all their rivers flow in flood, and many a hillside do the torrents furrow deeply, and down to the dark sea they rush headlong from the mountains with a mighty roar, and the tilled fields of men are wasted. . . .[8]

Justice, to the Greeks, was not simply fairness among people; it was keeping the proper relationships among people and between people, the natural environment, and the gods, so that the whole universe might stay in balance.

The balance was maintained, they held, not only by punishments but also by rewards. The goodwill of the gods was shown toward mankind through the favorable aspect of nature. Though the gods are often contemptuous and hostile, and their benevolence can be capricious, they respond with good gifts when people act in ways pleasing to them, and care for the natural environment wisely. The orchard and vineyard of a wise Greek farmer will be unusually fruitful because they are trimmed and cultivated with skill and care. "Such," Homer said, "were the glorious gifts of the gods."[9]

The way in which the natural environment, under the aegis of the gods, responds to care, good leadership, and justice is told beautifully in a simile from the *Odyssey* which is a perfect companion piece to the one from the *Iliad* quoted above. Here it must be remembered that Homeric kings went out into the fields and supervised agricultural labor—and not uncommonly they did it themselves—and that the usual epithet for a leader is "shepherd":

> Lady, no one of mortals upon the boundless earth could find fault with thee, for thy fame goes up to the broad heaven, as does the fame of some blameless king, who with the fear of the gods in his heart is lord over many mighty men, upholding justice; and the black earth bears wheat and barley, and the trees are laden with fruit, the flocks bring forth young unceasingly, and the sea yields fish, all from his good leading; and the people prosper under him.[10]

7

Greek Attitudes toward Nature

The ancient Greeks enjoyed nature. The dominant Greek attitude toward nature is delight in its many aspects. They were naturally encouraged in this regard by the beauty of the Greek landscape itself, which they recognized and to which they responded in its wild and cultivated aspects. The flashing sea, the rocky islands, the waving forests, all seen in the clear Greek light, made patterns which are reflected in their art and celebrated in their literature.

The freedom and grace of the creatures of the sea and land depicted in Minoan art was imitated by the Mycenaean ancestors of the Greeks, who added scenes of hunting and battle. The nonceramic painting of the Classical Greeks has almost entirely disappeared, but if descriptions and later Hellenistic and Roman adaptations are an indication, it included landscapes and animals in motion like those of the pebble mosaics at Pella, Alexander's Macedonian capital.

Greek architecture and sculpture included motifs from nature. Stylized leaves, flowers, and the heads of animals were repeated in architectural elements such as roof ornaments and the capitals of columns. While Greek sculptors were justly praised primarily for their portrayal of the human body, they also gave their careful attention to the natural world. Bulls and ravening lions, the elegant horses of the Parthenon, olive trees, and fountains all were carved in stone that still seems to have been given a life of its own. The tradition continued as long as there were Greek artists free to follow it. In the Hellenistic age, Artemis was accompanied by a vaulting deer, following a very ancient tradition that showed her flanked with wild animals as her attributes.

The poets also glorified nature. Homer, the first and greatest Greek epic writer, constantly uses epithets like "wine-dark sea," "Chalcis with its beautiful streams," "Pelion of the waving leaves," "Antheia with deep meadows,"[1] and countless others which are possible only for a writer who admires and responds to natural beauty and knows his reader does the same. Homer diverted his readers with many similes which depend on the power of nature's aspect to move the human mind:

> Even as in heaven about the gleaming moon the stars shine clear, when the air is windless, and forth to view appear all mountain peaks and high headlands and glades, and from heaven breaketh open the infinite air, and the shepherd joyeth in his heart. . . .[2]

In the *Odyssey*, after describing the beautiful natural setting of the nymph Calypso's cave, with its trees, flowers, birds, and springs of water, Homer says:

> There even an immortal, who chanced to come, might gaze and marvel, and delight his soul, and there the messenger Argeiphontes (Hermes) stood and marvelled. But when he had marvelled in his heart at all things, straightway he went into the wide cave. . . .[3]

Only a poet who admired nature's beauty would have one of his gods do so.

The lyric poets also sang of the pleasant aspects of land and sea. Sappho depicted a meadow pasture with spring flowers and light winds. While Archilochus made an unfavorable comparison between wild, forested Thasos and the lovely plains by the

River Siris, many passages in other poets indicate that the Greeks loved mountains and wilderness scenes as well as cultivated landscapes.

All the Greek dramatists used nature imagery, Euripides as much as any other. Among his fragments is the following:

> Dear is this light of the sun, and lovely to the eye is the placid ocean-flood, and the earth in the bloom of spring, and wide-spreading waters, and of many lovely sights might I speak the praises.[4]

Plato, whose philosophical writings contain many references to the natural environment, described mountains as both beautiful and useful.

A late school of pastoral writers at Alexandria, centered around Theocritus, adopted a romantic style of nature description which goes far beyond that of any other Greek writers in celebrating the beauties of the countryside. Theocritus has no rival in ancient times in his ability to communicate a sense of joy and delight in nature's beauty.

Beyond simply admiring nature, the Greeks tried to understand their environment. Their distinction from the other peoples of the ancient world consists in the fact that they tried to understand nature rationally, not mythically, and they did this in a sustained and purposeful way. Of course, myth and religion persisted in Greece and are often reflected in Greek philosophy. But the early philosophers seriously asked themselves questions about the substances which make up the world, and the processes which are going on in the world.

Thales, the first philosopher, came to the conclusion that all things are water. Others advanced air, fire, and earth as the basic elements, alone or in combination. Various sorts of motion were postulated as producing the changes which we see in nature, or the reality of motion was denied. All of these philosophers share a common Greek assumption about the natural world, that is, that it can be understood by the human mind because it possesses an inner rational order of its own. As Aristotle put it, "Nature does nothing in vain."[5] The natural environment, according to the Greeks, has unity and harmony in all its parts. In fact, as they soon realized, it looks as if it had been designed by a divine mind. Anaxagoras said that *nous,* or mind, was the first cause of the universe, and Plato expanded this into the idea of a single craftsman-deity who made the world.

The Greeks did not make a rigid distinction between living and nonliving things. Some believed the stars to be living creatures; others thought the soul was a material substance. Aristotle, for example, taught that the living and nonliving merged one into another gradually. The whole universe could thus be conceived by the Greeks as a living, breathing organism. All life, sharing the same substance and forming part of the living world, therefore, shared a certain sympathy of affinity.

It is, of course, impossible to speak of only one Greek attitude toward the natural environment. Among these keen and argumentative people, each idea seemed to generate its opposite, and there were other philosophers who denied much of what has just been said.

Leucippus and Democritus believed that the world is purely physical, being composed of indivisible particles called atoms, whose movements are purely mechanical and governed by accident. This view denies the idea of design and purpose in the universe, although it still bears the marks of rational consistency so characteristic of Greek thought. According to Epicurus, whose philosophy follows these lines, there is no creator or designer other than nature itself, and nature works through blind physical cause.

Whether through design or accident, the Greeks believed, the environment has great influence over mankind. The latitude and climate, they felt, determined the size, strength, and other characteristics of the inhabitants of a country. People like the

Ethiopians, who live near the places where the sun rises and sets, are black; and those who live in places where the air is thick will have slow and phlegmatic temperaments. The physician Hippocrates, father of medicine, is perhaps the author of a work entitled "Airs, Waters, and Places," which points out the importance of environment in the cause, diagnosis, and treatment of illnesses. By knowing the climate, exposure, and quality of the water in a place, Hippocrates taught, a physician could know what diseases to expect among the people living there, and could suggest changes which might help to heal his patients.

Other authors stressed the influence of environment on human history. Thucydides suggested that the thin, dry soil of Attica made that land unattractive to potential invaders and had saved it from conquest. Democritus believed that many of the advances made in human civilization are the result of observing the habits of other animals. People have learned how to weave from the spider and how to sing from the birds. They build houses of clay because they have watched the swallow at work.

But mankind to the Greeks was not a mere victim or pupil of the environment. They saw the human species as able to alter the world as no other creature can. Mankind is apart from the animals in ability to reason and take forethought. Sophocles, in his play *Antigone,* gives the chorus a great hymn to sing in praise of the ability of mankind to control and change the earth and its creatures, although it ends with an ironic twist. Man, the chorus sings, can cross the sea and plow the earth. He snares birds and beasts, and tames the horse and mountain bull. He knows speech and thought, and how to escape frost and rain, but not how to escape death, or to prefer justice to evil.[6]

It was the opinion of Anaxagoras that human beings were cleverer than the beasts because they had hands with which to manipulate. The Greeks seem to have realized that mankind was less advanced in past ages, and attributed civilization to two things, the use of fire and the cultivation of grain. They knew that agriculture had been practiced in Greece for many generations, and that they looked upon a landscape which had been much altered by human activities. Greek writers from Hesiod on were fascinated by agriculture, and believed that through it, mankind was changing the earth by creating ordered patterns of beauty. Strabo, a late Greek geographer, believed that people worked in partnership with nature, to rectify the deficiencies of the environment. Egypt, he believed, was not so much the gift of the Nile as the achievement of the labors of the Egyptians. Greeks could admire a fine wild scene, but only the richness and regularity of a cultivated landscape called forth their unstinting praise.

The Greeks were not always optimistic about changes wrought in nature, however. Herodotus, as noted above, felt that many mighty works, like bridges and canals, were dangerous infringements of the natural order. In one of the best analyses in ancient times of human impact on the earth, Plato described the deforestation of Attica and the resultant soil erosion and drying of springs, so that "what now remains compared with what then existed is like the skeleton of a sick man, all the fat and soft earth having wasted away, and only the bare framework of the land being left."[7] The conclusion that the earth under mankind's hand is undergoing degeneracy, not progress, was reached by many Greeks, and was reinforced by their memory of the old legend of the Golden Age told by Hesiod. In the Golden Age, the earth produced fruit and grain by itself, without the need of cultivation. All beasts and birds were friendly and helped human beings out of their own free choice. In succeeding ages, this idyllic state of affairs altered; labor and strife became the human lot.

Change was not regarded as an automatic good by the Greeks. They preferred stability, and were suspicious of alteration. It is therefore not surprising to find that as the large cosmopolitan cities of the Hellenistic age replaced the small-town poleis (city-states) of Greece, Greek writers began to stress the superior virtues of the older agricultural life, when even town dwellers could have farms within walking distance, and people were closer to the land. For a charming and explicit development of this idea, Dion Chrysostom's *Euboean Discourse* is well worth reading. After describing the idyllic life of simple hunters and shepherds isolated in the back country, where they lived happily in natural honesty, hospitality, and unspoiled nobility, the author brings them into the city and into contrast with the corruption of "civilized" society. He makes it clear that he considers urban life as a state which breaks down human character.

Perhaps the most important Greek attitude toward nature was a certain curiosity which, combined with a willingness to depend on reason, led to the first great strides of science. True enough, the Egyptians and Mesopotamians had already amassed a great deal of practical and theoretical knowledge in the fields of medicine, astronomy, and mathematics. The Greeks expanded upon this in a relatively short time, following the spirit of Plato's remark that explanations need to account for what is actually observed. Of course, many of their answers now appear to have been wrong, and, more important, the methods they used to arrive at answers were inadequate. They asked many of the right questions of the natural environment, however, and as scientists they had no rivals in the ancient world.

The Greeks did not consciously invent the science of ecology. In spite of the fact that it comes from good Greek roots, the word *ecology* was not used before the nineteenth century. But the philosophers did ask questions concerning the relationships of various living things, including people, to one another and their environment. These questions could be called ecological, and in answering them, the Greeks came to recognize some ecological principles.

Anaximander, a friend and pupil of Thales, was the first

thinker whose speculation on ecological questions has survived. In discussing the origin of mankind, Anaximander wondered how human beings, who spend a very long childhood in a defenseless state and who are in any case much weaker than many other animals, could have survived in the earliest times. He believed the answer was that they had originally grown as embryos inside creatures like fishes, in which form they were better protected from predators.

Empedocles added a rudimentary form of the idea of natural selection. Believing that all creatures arose from a random concatenation of the elements, he stated that only those whose structure fitted their purpose had actually survived. Those which found themselves with an odd assortment of parts perished.

Herodotus, whose interest in natural history was wide-ranging and who sometimes repeated fantastic stories about animals and plants without necessarily believing them, also pondered the problem of the relationship between predators and prey. He noticed that timid animals which are eaten by others produce young in great abundance, while the predators bring forth only a few offspring. Thus a balance of numbers is achieved. This idea, now recognized as a basic ecological principle, was repeated by Plato, who put it into the mouth of Protagoras in his dialogue of the same name. Protagoras is made to say that the gods have given fur, claws, wings, and the like to animals so as to compensate them with defenses against one another.[8] This is an early statement of the idea of the balance of species.

Aristotle is the most important Greek writer in the field of biology, and his influence in later times is almost incalculable. He is noted for his careful classification and description of animals. Some of his statements, once thought to be fanciful or inaccurate, have now been found to be true. For example, he said that the catfish guards its eggs. Most European catfish do not, but recent observation has shown that Greek catfish do. Many of Aristotle's descriptions are based on specimens sent back to him from Asia by his ex-pupil, Alexander the Great, and reports prepared by the philosophers who accompanied the conqueror.

Aristotle's teachings about animals are the foundation of much of Western thinking about the relationship of mankind to the whole natural environment. His reasoning was teleological. All things have a purpose or end for which they are formed. When a thing fulfills its end, it is useful and beautiful. Therefore, no animal lacks beauty, because all animals are formed for their proper ends. And what is their proper end? Aristotle says it is the service of mankind. All animals, and indeed all other things, exist for human good. Therefore they are fit instruments for human beings to use, as in domestication. Of course, Aristotle himself would not have justified the misuse of animals or their senseless slaughter, but once he became established as an authority, his teaching that other creatures are of a lower order, subservient to human needs, bore with it the obvious corollary that they have no justifying purpose of their own, and therefore no independent right to existence.

Aristotle's student Theophrastus has long stood in the shadow of his teacher, and has thus been underestimated. He has often been called the Father of Botany, but a study of his writings from the standpoint of ecology shows something of his true greatness. He deserves another title: Father of Ecology. More than half of Theophrastus's botanical writings deal with ecological observations. It is not a matter of isolated passages, but of a consistent viewpoint. Theophrastus did not view a plant in isolation, but asked what its relationship was as a living organism to sunshine and exposure, soil and climate, water and cultivation, and other plants and animals.

He based his statements on actual observation in many cases. He sometimes mentioned a tree which he had seen and measured, such as the plane tree by the watercourse in the Lyceum at Athens. Like Aristotle, he had Alexander's specimens and reports to use.

Theophrastus did not accept Aristotle's idea that all animals and plants exist to serve mankind. He did not deny that there is purpose in nature; he found the purpose of an annual plant, for example, to be reached in the production of fruit and seed, thus providing for a new generation. But the purpose of things in nature, he maintained, is not always evident. What is the purpose of droughts and floods? He asked for an "effort to determine the conditions on which real things depend and the relations in which they stand to one another" through careful observation rather than the facile assigning of final causes.[9]

Nothing grows or flowers before its proper season. Various plants prefer different climates, and the attempts of people to grow them in countries far from their point of origin often fail. A fruit tree planted in another country may flower but not set fruit. Theophrastus recognizes not only the difference between major climate zones, but also microclimates. Some trees grow well on the sunny slopes of mountains, others on the shady northern side. Plants which are fond of water will not grow well in dry, sandy locations, and the reverse is also true. So strong are these preferences as to locality that after a devastating flood, the same kinds of trees will come up in the same places, as if following the example of those that grew there before. Those who plant seeds should consider the soil, sun, and winds, because "locality is more important than cultivation and care."[10] The habit of growth of plants will vary under changes in these conditions. He also observed that mountains offer a wide variety of conditions because of elevation and aspect, and he knew of the existence of narrow endemics particularly on isolated mountains. He is acquainted with the extreme importance of water to plant growth in arid regions.

Theophrastus takes a particular interest in the response of plants to domestication. Some plants, he says, cannot be cultivated, but of those which can, some take on a very different appearance under cultivation owing to the soil, irrigation, and manure which they receive. He discusses the changes produced by cultivation, such as the tendencies to produce fewer fruit of better quality and to grow straighter with fewer knots. The habit of growth depends on the distance between plants. He discusses various weeds and their means of spreading.

Finally, Theophrastus notices certain local changes in climate brought about human activities. His teacher, Aristotle, had postulated the existence of long-term, major changes in climate caused by the aspects of the heavens, including a "great winter" many years in length, but Theophrastus gathered information on actual temperature changes during his own time in Greece caused by the draining of marshes, the alteration of the course of a river, and deforestation.

Greek science received a significant impetus from the Museum, a research institute in Alexandria. When Alexander the Great conquered Egypt, he founded the city as his capital and named it after himself. On Alexander's death, Ptolemy, one of his generals, became king of Egypt. He and his successor, also named Ptolemy, invited Demetrius of Phaleron and Strato of Lampsacus, both former students of Aristotle and Theophrastus, to Alexandria to found and help direct a great scientific, literary, and religious arm of the palace, under royal sponsorship and scrutiny. It was not so much a school as a place where philosophers and poets could pursue their studies and respond to the commands of the king.

The Museum, so called because it was dedicated to the Muses, the goddesses of the various fields of literature and science, included the greatest library of the ancient world, containing almost every known book in Greek. Connected with the Museum was a botanical garden with trees and plants from many parts of the world, and a zoo with a large collection of animals and birds, even including a polar bear, and all these were available for study by the scientists in residence.

Important discoveries about the natural environment were

made at Alexandria. In astronomy, both the geocentric and heliocentric theories were elaborated. In geography, the size of the earth was accurately measured and the existence of the climatic belts discussed. In medicine, anatomy was studied through careful dissections. There and elsewhere in the Hellenistic world, botanical manuals were written and illustrated with drawings. Unfortunately, many of the writings of the Museum's scientists have perished. The Museum of Alexandria represented a focal point in the Greek study of the natural world, and bridged the time between the great Greek philosophers of the fourth century B.C. and the rise of Rome to dominance in the Mediterranean Basin.

8

The Impact of Greek Civilization on the Natural Environment

The Greeks severely altered the ecosystem in which they lived, exhausting important natural resources and contributing in some ways to their own decline. The landscape which the world knows today as typically Greek is to a large extent the result of human actions in ancient times. Of course the Greeks did not find their land a wilderness, since agriculture had already been practiced there for three or four thousand years when they arrived, and after the period of Classical Greece and the Roman Empire, many of the same processes of human use and change continued.

One of the most striking effects on nature in Greece was the removal of the forests, and this occurred primarily in the Classical and Hellenistic periods between 600 and 200 B.C., except for the more remote mountainous areas. Trees were used for fuel at all times, the wood being burned, or first reduced to charcoal and then used in the firing of pottery and in various industrial processes, especially the reduction of ore from the mines. Wood was necessary for all sorts of construction, since even marble temples had roofs supported by wooden rafters. Carts and chariots, doors and furniture, handles for weapons, and objects of art were made from various kinds of wood. Pitch, tar, and resins used for waterproofing were sometimes extracted from trees by fire. Greek wine was usually flavored and preserved by adding pine resin, collected from live, scarified trees, which eventually died. As the Greek cities began building navies in earnest, the demand for fine lumber and tall trees for ships' masts greatly increased, and it became necessary to import these products from forested areas at greater distances.

The iron axes of Greek loggers made the forests resound, according to a famous description in Homer, and no tree was too lofty or too great in girth to escape them. Logs were floated down the rivers, and the lumber was exported by sea. But the activities of lumbermen and charcoal burners were not the only forces destroying the Greek forests. Forest fires, many of them set on purpose by shepherds, raged unchecked. And natural regeneration of the forests, which is slow in the dry Mediterranean climate, was usually prevented by the practices of grazing. Shepherds often girdled trees to improve the grass underneath. Swine were pastured in oak forests. Even ordinarily grass-eating animals like sheep and cattle are forced to eat leaves and twigs in the Greek hills, and goats, which browse on bushes and small trees by preference, will also climb into large trees and eat the foliage. Goats made permanent the deforestation of thousands of square miles of Mediterranean hillsides by eating every seedling tree that ventured to show its head, until there were no more left. The removal of the forest habitat, along with the intense competition of grazing animals, reduced the numbers of wildlife.

Once the land was bare of trees, the torrential rains of the Mediterranean fall, winter, and spring washed away the unprotected earth. Unimpeded erosion destroyed the uplands that might have grown trees again, and the silt, sand, and gravel which reddened the rivers was deposited at their mouths along the shores of the virtually tideless Mediterranean Sea. There, new swamplands extended for miles and served as breeding grounds for mosquitoes. Malaria is believed to have entered

Greece in the fourth century B.C. Marshes are useful as homes for water birds and other wildlife, and can be reclaimed for agriculture, but such reclamation had to wait for centuries.

Perhaps half the land area of Greece was originally covered by forests; today they occupy less than one-tenth. The ancient Greek writers were very much aware that deforestation was going on, and that it was the result of human actions. Islands and districts described by Homer as "wooded" were desolate by the fifth century. Thucydides, who also speculated on causes of forest fires, recognized that the demands of shipbuilding had exhausted the forests in some areas. Plato knew that Attica had suffered deforestation in the space of two generations, and gave concrete evidence of this fact in the *Critias,* which is worth quoting here:

> What now remains compared with what then existed is like the skeleton of a sick man, all the fat and soft earth having wasted away, and only the bare framework of the land being left. But at that epoch the country was unimpaired, and for its mountains it had high arable hills, and in place of the "moorlands," as they are now called, it contained plains full of rich soil; and it had much forest-land in its mountains, of which there are visible signs even to this day; for there are some mountains which now have nothing but food for bees, but they had trees not very long ago, and the rafters from those felled there to roof the largest buildings are still sound. And besides, there were many lofty trees of cultivated species; and it produced boundless pasturage for flocks. Moreover, it was enriched by the yearly rains from Zeus, which were not lost to it, as now, by flowing from the bare land into the sea; but the soil it had was deep, and therein it received the water, storing it up in the retentive loamy soil; and by drawing off into the hollows from the heights the water

> that was there absorbed, it provided all the various districts with abundant supplies of springwaters and streams, whereof the shrines which still remain even now, at the spots where the fountains formerly existed, are signs which testify that our present description of the land is true.[1]

Plato based this passage on actual observation, as evidenced by his reference to the rafters and the spring-shrines which could be seen in his own day. In it, he shows his understanding of the role of forests in conserving water, and the serious effects of their loss.

By Theophrastus's time, the forests of Mount Ida, on the coast of Asia Minor, were almost exhausted, and most good timber came to Greece from Macedonia and Chalcidice, where some mountains are still forested even today, and from as far away as the Black Sea coasts. Pausanias, a travel writer of the second century A.D. in Greece, mentioned trees wherever he saw them, which shows that they were rare enough to merit comment. The general pattern of deforestation is the removal of the woods first near settlements and in the lowlands, then from the mountain slopes; finally the only forests remaining were in high and distant places.

Forestry in Greece soon became the concern of the government of each polis, due mainly to the need of lumber for ships. Inscriptions show that Greek states controlled the cutting of timber on their own territory and required replanting in some cases, which approaches a policy of raising forest trees by cultivation. The supply of wood from abroad became the subject of international negotiations, as in the case of the aid given by a Persian satrap to the Spartans during the Peloponnesian War. The Athenian expedition to Sicily in the same war was partially to secure timber for the navy. Cities controlled the import and export of lumber through a system of licenses, which proved quite lucrative for some of those who obtained them. Colonization in forested areas like Chalcidice was undertaken with the idea of developing a lumber trade by the mother city. Other Greek colonies found the forests a barrier, and undertook deliberate clearing.

A good example of changing governmental forest policy occurred in Cyprus, where in the fourth century B.C., according to Theophrastus, the kings carefully conserved the trees, but in the next century, Eratosthenes reported, the rulers offered encouragement to those who cut the forests, even so far as to offer them title to the cleared land.[2]

Hunting, like forestry, uses living things which can be replaced by natural reproduction, but animals, like trees, were used beyond the possibility of renewal. There were some professional hunters in Greece, and wild meat and skins made a small contribution to the Greek economy. Furs from animals like the beaver, which was trapped in northern Greece, were used for warm clothing and decoration in the backcountry. Homer often depicted his heroes dressed in animal skins. Hunting was a sport for the nobility of Greece, whose exploits in pursuit of wild boar and other creatures are celebrated in literature. More animals and birds were killed to protect domestic animals and the growing crops; the Homeric simile of a lion pursued by herdsmen is well known. Large numbers of smaller game were simply killed for food; many songbirds which would be regarded as purely ornamental today in America were, and indeed still are, killed and eaten in Greece.

The larger predators were decimated. The lion and leopard were extirpated from Greece and coastal Asia Minor by the end of the Hellenistic Age. Wolves and jackals were rarely seen outside the mountains. Hunting reduced wild cattle, sheep, and goats to small remnant herds, and completely eliminated them from some of the islands. Other animals and birds suffered, not just from hunting, but also by the modification of their habitats by the spread of agriculture, grazing, and deforestation. Of course, some forms of wildlife are adapted to live in disturbed areas or in close proximity to people, and these prospered, but the total effect of ancient pressure upon wildlife was the extinction of some species, the introduction of others, and the general alteration of the ecosystem.

The heroes of Homer were meat eaters, but by the time of the Golden Age of Greece, meat was rarely eaten except on religious days of sacrifice, and seafood was the major source of protein in the Greek diet. Entire villages consisted of fisherfolk, whose work was to supply Athens and the other cities with fresh, salted, or smoked fish and shellfish, along with squid, octopuses, and other delicacies. Divers brought up sponges, which had many household and industrial uses. The famous murex, or purple shells, the source of the reddish dye in royal and aristocratic robes, was a Phoenician monopoly for a time, but some of the best murex fisheries were in Aegean waters. The purple shells were taken so intensively that the supply was exhausted in some areas. The extent of the depletion suffered by other fisheries is unknown, except that fishermen seem to have ventured farther and farther from their home ports. The Mediterranean continued to be a good source of fish until recently.

Mining and quarrying had a widespread and noticeable effect on the Greek landscape. Herodotus remarked that gold mining on Thasos had thrown a whole mountain upside down, and scars left by operations of this kind are still visible today, although many of them are being obliterated rapidly by larger modern projects. The ancient Greeks were industrious miners, and in many cases completely worked out bodies of ore at the level of technology of which they were capable, a level which was surprisingly high and efficient, considering its date. They mined in underground tunnels and open pits, and separated gold from alluvium. Their tools were of types closely resembling those still in use, such as shovels, hoes, picks, hammers, chisels, and crowbars, and were made of iron or a type of steel. Machines such as the Archimedean screw pump and the water-wheel aided in some mining operations. Work in the mines was the hardest and cruelest known to ancient people, so almost all of it was done by slaves and condemned criminals. Their use was governed by contracts between the states and wealthy citizens who guaranteed their labor and profited from it. The workers had to crawl through poorly ventilated tunnels seldom more than three feet high, dragging heavy sacks of ore; their life expectancy was not long. No one knows how many people were affected by poisonous substances, such as lead and mercury which were associated with more valuable ores in the mine, or by drainage of toxic minerals from the mines into rivers and other sources of water, but many of the pigments, paints, and glazes used by the Greeks were poisons.

Gold, silver, and their natural alloy, electrum, the most valued metals, were almost completely mined out in Greece before the end of the Hellenistic period. The most famous mines in Greece were the Athenian silver mines at Laurium, whose operation is known in detail, and the Macedonian gold mines of Mount Pangaeus, developed by King Philip, Alexander's father. Copper was obtained from places where it was abundant, particularly Cyprus, and tin, the other constituent of bronze, was always rare and had to be shipped from even more distant spots such as the Tin Islands (Britain) and Spain. Iron, fairly plentiful in Greece, was not exhausted, but especially rich ore bodies like the Spartan iron mine in southern Laconia were intensively mined. Lead, asbestos, various precious stones, and asphalt were mined in Greece, but amber was imported from far to the north. The properties of coal were known to the Greeks, and it is not rare in their homeland, but they seldom used it as a fuel, so the deposits were not much mined. Salt mines were also few, since most salt was evaporated from seawater.

motive might have been felt by the first generation or two of invaders in the country, who came from the mountains, and the persistence of the same forms among their descendants could be explained by the almost universal conservatism of religious architecture. The mountainlike appearance of ziggurats and palaces in Mesopotamia was undoubtedly emphasized by the trees, shrubs, and vines which were planted on them. Archaeologists have noticed that ziggurats were provided with interior channels to drain the water which seeped down from the carefully watered plants on top. So striking were they in appearance that they were among the wonders of the ancient world, the Hanging Gardens of Babylon. Kings enjoyed collecting plants and animals from distant parts of the world to add interest to their gardens, which became botanical and zoological parks as a result. Some things discovered in this way may well have been adapted to agriculture; perhaps the grapevine and olive tree first came to Mesopotamia as gifts of specimens to a royal garden. Domestication seems to have been a key idea in royal collecting. One king had a pet lion who was supposed to have fought beside him in battle against his enemies.

The cities of Mesopotamia have been desolate mounds for a score of centuries, and only a poor remnant of the "Fertile Crescent," that green, cultivated area which once arched across the Middle East from Sumeria to Palestine, is visible in photographs taken from space today. This disaster is due not simply to changing climate or the devastating influences of war, though both of these have had important effects. It is a true ecological disaster, due partly to the difficulty of maintaining the canals and keeping them free of silt, but more importantly to the accumulation of salt in the soil. Irrigation water, carried over large areas, was allowed to evaporate with insufficient drainage, and over the centuries in this land of low humidity and scanty rain, the salts carried in by the water concentrated. Such areas had to be abandoned, while new sections were brought under

Quarrying was a major activity in ancient Greece. Marble, limestone, conglomerate, and, less often, granite for the construction of temples and other public buildings came from well-established sources in the islands and the mainland mountains. Mount Pentelicus, the source of pentelic marble for the Parthenon and many other buildings in Athens and elsewhere, bore a huge gleaming white scar on its side. Using metal saws and wooden wedges which expanded when they were soaked in water, the workers could remove huge blocks of the pure white stone. These were roughly dressed at the quarry, packed in wooden crates, and transported on sledges and carts to the building site, where they received the final desired shape. Quarries became artificial chasms in places like the island of Paros, whose glistening white marble was justly famous, and Syracuse, where the excavation became a concentration camp for Athenian prisoners during the Peloponnesian War. A special type of quarry of great importance in Greece supplied the fine clays used by potters for cups, pitchers, amphorae, and countless other utensils and vessels. The scars of excavation and subsequent erosion produced by mining and quarrying in Greece, although scattered, never entirely healed over with a mantle of vegetation and soil.

The most consistent and widespread force of environmental degradation in ancient Greece was the grazing and browsing of domestic animals. Four-fifths of the land area, unsuited to cultivation, was used as pasture. Prevented by mountain barriers and political boundaries from practicing nomadism, the Greek herders utilized narrowly limited areas and in virtually every case subjected the vegetation to severe overgrazing. They moved the herds every year from the lowlands in winter to the mountains as the new growth appeared in spring, and back again in fall. The commonest animals, sheep and goats, were the most destructive but also the most useful, since they supplied meat, milk, and the leather and wool which served as the raw materials for almost all clothing. Sheep eat grass and other plants right down to the roots, while their sharp hooves churn up the soil. Goats, preferring to browse on the tender shoots and foliage of shrubs and trees, will eat almost any available vegetable material if need arises. Together, these two disparate animal types can strip a hillside bare, opening it to erosion, driving away competing wildlife, and forcing the whole ecosystem to regress down the scale of succession and energy. Wise limitation of numbers could have prevented this, but all the evidence indicates that such limitation was almost never practiced. If one herder left any vegetation unused by his flocks, it would probably have been used the same season by others.

Cattle, less destructive than the smaller animals, could still overgraze and cause less desirable plants to increase by selectively eating the most desirable ones which compete with them. Cattle in Greece were limited to sections possessing rich lowlands, like Boeotia, Elis, and Messenia, though they were often pastured in the mountains during the summer. Pigs were characteristically driven into oak forests to eat the acorns. Many Greek leaders went to considerable expense to import fine stock from distant places. Polycrates of Samos, for example, brought in improved breeds of sheep and pigs.

Horses in ancient Greece were indelibly associated with war, and with the aristocrats who alone could afford them. Aristocratic names in Greece often contained the word *hippos* ("horse"), such as Hippodamos, Pheidippides, Xanthippos, and Philip (Philippos). The terrain and vegetation of Greece were suited neither to the raising of horses nor to their use in battle, with the exception of the few large plains like those in Thessaly and Macedonia. Far more common was the tough little Mediterranean donkey, better able to use the meager vegetation and scarce water of the Greek highlands, and its mixed offspring, the mule, which combined the greater size of the horse with the stamina of the donkey. These animals were the major source of land transportation, and had to be fed by hand for much of the year.

Agriculture was the main economic activity of ancient Greece, and many Greek authors commented on this phase of the relationship of mankind to the environment. They realized that there was a time when people did not cultivate the earth, and that farming represents a constant effort of mankind to alter the earth and make it more fruitful for human purposes. The general pattern of Greek agricultural land use was that of small farms, cultivated intensively by owners who lived in nearby villages. Isolated farmhouses were rare, due to danger from hostile raids and the scarcity of water. Greek farming villages were, and still are, generally built on outcroppings of rock or other unproductive land to leave all the precious soil free for cultivation. A few large farms, often devoted to one productive crop, like olives, and worked by slaves, existed and became more usual in the Hellenistic period.

The characteristic Greek crops, as elsewhere in the Mediterranean, were grain—chiefly barley and wheat—olives, figs, grapes, various other fruits and vegetables, and hay. With the pattern of mild, wet winters and hot, dry summers, these crops could keep farmers busy almost all year, with a slack season in the late summer before the vintage. They plowed for grain in the autumn with oxen, using a simple plow with an iron share, planting wheat with the rains of October or November and harvesting it with a sickle in May or June. Grapevines and olive and fig trees bore fruit in the fall and early winter. Vegetables and fodder matured in the early summer.

In addition to these crops, the Greek farmers kept bees, as honey was to the ancients what sugar is to modern Americans. After their introduction into Greece about 700 B.C., chickens were raised, as well as geese and other poultry. A few people on the island of Cos discovered how to gather and weave a rough silk from the native silkworm.

Greece is a rocky country, not blessed with many rich soils, but with proper treatment, the soil continues to bear good crops generation after generation. Through trial and error, the Greek farmer had learned how to judge the soil and how to treat it, and whenever possible used good conservation methods. The Greeks knew the use of manure and compost, but were handicapped because the manure of grazing animals was not available for half the year. (The temple at Delos made a business of exporting pigeon guano.) They had discovered the value of letting the land lie fallow and of repeated plowing. They knew that legumes enriched the soil, and grasped the principle of crop rotation. Unfortunately, the small size of Greek farms and the necessity of planting crops only on the soil and exposure which would grow them best precluded much use of rotation. Green crops were plowed under as fertilizer as early as 400 B.C. Mineral fertilizers such as lime and nitrates were sometimes spread on the land. At Aegina, where a thin layer of infertile soil and travertine occurred over a rich marl, the marl was excavated through holes in the surface and spread over the land. The resulting caves were used for storage, or even for dwellings, which caused the Aeginetans to receive the name of Myrmidons, or Ants.

Contour plowing and terracing with stone walls to slow erosion were widely practiced. Even today, one can watch Greek farmers carrying soil back up onto the terraces from which it was washed by the rain. Much of the hillside land they reserved for orchards and vineyards. They used irrigation, which was absolutely necessary for summer crops, and preferred wells and springs as sources, since streams and canals carry weed seeds. In any case, Greece is a country without large rivers, and no great irrigation works like those in Egypt and Mesopotamia were constructed. In much of Greece, irrigation was only possible for gardens, and the preponderance of acreage was dry-farmed. Water rights were controlled by strict and detailed customary laws.

That the Greeks knew how to treat the soil and guard its productivity is shown by Xenophon's remark that the best way for a prudent man to make money is to buy up deteriorated farmland, renew it by careful husbandry, and sell it at a better price. That they were not always able to apply their knowledge is also clear.

Population pressure on the available land caused the great colonial expansion of the Greeks from the eighth to the fifth centuries B.C. The Greeks cultivated not only the land best suited to cultivation, but the marginal stretches as well. Sparta, determined to retain an exclusively agricultural base for her military establishment, expanded her land by conquering the large, rich neighboring state of Messenia and using the Messenians as agricultural laborers. Athens took another route, becoming a great commercial sea power, but the reforms of Solon which laid the foundation for that development presupposed a land base inadequate to feed the people of Athens. Solon saw that Athens's deteriorating soil could not raise enough wheat, but could support olive trees sufficient to produce a great surplus for export. His cancellation of agricultural debts and his guarantees for the freedom of all citizens were meant to strengthen the class of independent farmers, and the subsequent agricultural loans of the tyrant Pisistratus, combined with the seizure and division of large estates, were aimed at the same goal. But Solon and Pisistratus effectively turned Athens's agriculture into a virtual one-crop system for export, and that crop, olives, was vulnerable to the attacks of their enemies' armies.

The part taken by the Greek governments in agriculture is quite evident; it was important and of long standing. In colonies, land was apportioned by the city government. Laws governed the sale and transfer of agricultural land, usually to discourage its alienation. The larger royal governments of the Hellenistic period were able to accomplish even more. Swamps were drained, as was the great Boeotian Lake Copais, once a source of eels and fish for Greek tables but since its drainage by one of Alexander's generals a flat, rich farmland. Hellenistic kings encouraged agricultural research, including attempts to acclimatize plants and animals from other lands. Greek engineers at Alexandria, Syracuse, and elsewhere invented more efficient machines, including pumps and wine and oil presses. The water-driven mill for grain was widely used from the second century B.C. on.

Greek agriculture did not exhaust the soil, although erosion was a great and continuing problem. But Greek farmers, forced to squeeze every possible use out of a severely limited area of arable land, did see that land deteriorate. In some areas of Greece, two crops a year were forced from the soil, and in southern areas like Crete, even three. Despite the care and knowledge which have always characterized Greek farming, much of Greece was practicing subsistence agriculture, and the biotic community which constitutes productive soil suffered.

The polis, or city, was the dominant social, political, and economic institution of the Greeks, and hundreds of poleis were organized, not only in Greece proper but also on most of the coasts of the Mediterranean and Black seas where no one stopped them. Cities obviously alter the natural environment. Not only do they themselves take up land and constitute an artificial environment, but every city affects a hinterland of some extent through its demands for food and other products and raw materials. Many of the extractive industries mentioned above existed in large measure to fulfill the needs of the cities.

In early and Classical Greece, the cities remained small and, in spite of the development of manufacturing and commerce, basically agrarian. Most citizens in almost all cities had farms in the adjacent countryside. Athens, one of the largest Greek cities, had a citizen population of about 30,000 adult males, many of whom lived in villages outside the city walls. Perhaps 100,000 people lived inside the walls at one time, and the total population of Attica was probably between 250,000 and 400,000. Much of the harder labor of the city was done by the numerous slaves, and resident foreigners carried on a large proportion, but by no means all, of the commerce.

The Greeks tended to assign a lower social position to those who labored with their hands, an attitude which seems associated with the existence of slavery. The idea of a growing economy and the need for expanding exploitation of natural resources was foreign to them; the Greek desired a stable, secure state independent of others in economic as well as political respects.

Trade and colonization developed to a considerable extent because of the agricultural facts mentioned above. Many Greek cities did not find their adjacent farmlands adequate to supply them with grain, so they had either to engage in trade with productive lands like Egypt and Scythia (southern Russia), or to send out bands of their citizens to found new cities where land was available, in the Northern Aegean, the Black Sea coasts, Sicily, Italy, and elsewhere.

These activities of the Greeks, trade and colonization, affected the natural environment in various ways, by spreading domestic animals and plants such as the vine and olive, creating new demands, developing new farm and grazing lands, cutting forests, exploiting mines, and in particular by founding cities which were new, self-conscious creations and not mere dependencies of the mother cities.

The old Greek cities had grown haphazardly. Centered around a fortified height, or acropolis, which may have existed since Mycenaean times or earlier, many of them were located a few miles from the sea to discourage pirates. The streets in a city like Athens were a planless jumble of narrow passageways pierced only by ceremonial roads such as the Sacred Way. The center of civic life was the marketplace, which had facilities for the city government as well as trade. Not too far away, four miles in the case of Athens, was the port, which expanded with the growing importance of commerce.

It was in new colonies, expanding ports, and the refounding of older cities smashed by war that Greek city planning found its sphere. Here the Greeks could consciously create an environment in which to live, and they did so. Hippodamus of Miletus, the famous city planner and architect, is associated with the rebuilding of his native city after the Persian Wars, a new plan for Athens's port city of Piraeus, and the foundation of Thurii, an Athenian colony in southern Italy. He favored a regular plan of straight streets crossing one another at right angles to make rectangular blocks, some of which were set aside for public buildings and the marketplace. This pattern was adopted by most later Greek city planners and, with modifications, by the Romans and in most American cities. Hippodamian streets were narrow, however, and the planner made no provision for broad avenues or impressive architectural views. His pattern has a rational rigidity which presented problems for towns built on hills, with irregular city walls. Aristotle knew the Hippodamian plan and liked it, but he favored irregular streets in parts of the city to confuse invading enemies. Cities built around harbors often took their street plans from the arrangement of aisles in a Greek theater. A city being rebuilt on an older site needed to take account of the location of temples and sacred precincts which would not be disturbed. Hippocrates, with an eye to the healthiness of cities, advised that their exposure should take account of the sun and the prevailing winds. As a general principle, he believed a city should face eastward. He also urged care in the siting of individual houses in relation to sun and wind. Most Greek houses were small and simple, but whenever possible they were centered around an inner court, and the rooms opened into this rather than outward, a good plan for a warm, sunny climate like that of Greece.

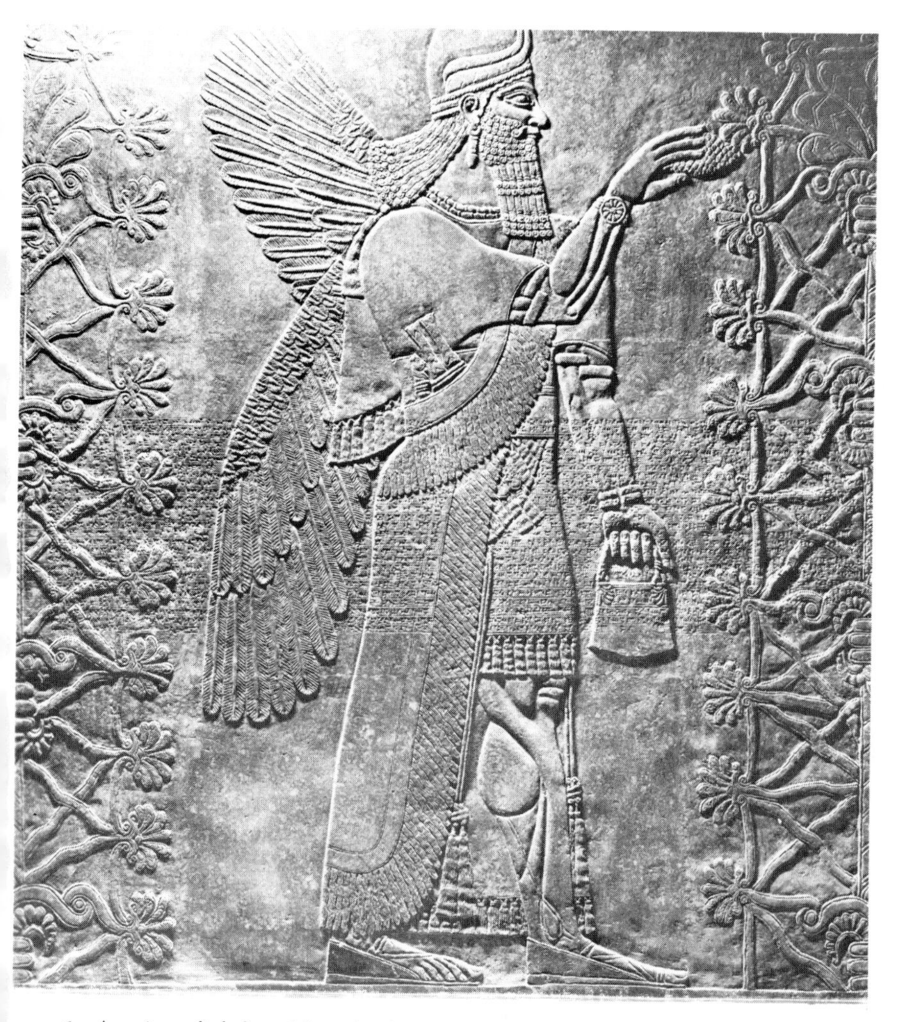

1. Assyrian relief, from Nimrud, 9th century B.C. A winged Assyrian deity is shown fertilizing a stylized date palm by applying pollen, a common agricultural practice in Mesopotamia. The relief implies divine sanction for the orderly round of seasonal agriculture. *Nelson Gallery, Atkins Museum of Fine Arts, Kansas City, Mo.*

2. Egyptian figure of a hippopotamus, faïence, from Tomb of Senbi, at Meir, 12th Dynasty. Careful observation and sympathetic portrayal of the animal are evident. The environment of the hippopotamus is represented by the drawings of water plants on its surface. *The Metropolitan Museum of Art, Gift of Edward S. Harkness, 1917*

3. Egyptian figure of a gazelle, ivory, 18th Dynasty, from Thebes. The gazelle belongs to a desert ecosystem as the hippopotamus belongs to a riverine ecosystem, and the base of the figure is painted to depict a desert crag, with desert plants. *The Metropolitan Museum of Art, Carnarvon Collection, Gift of Edward S. Harkness, 1926*

4. Egyptian wall painting: the vintage (copy). From Tomb of Apuy, the Sculptor, Thebes, 19th Dynasty. Workers at left are crushing the grapes with their feet while holding ropes to prevent them from slipping and falling. Figures at right are gathering grapes. The large rosettes represent vine leaves. The vine was a cultivated plant introduced into Egypt in early times. *The Metropolitan Museum of Art*

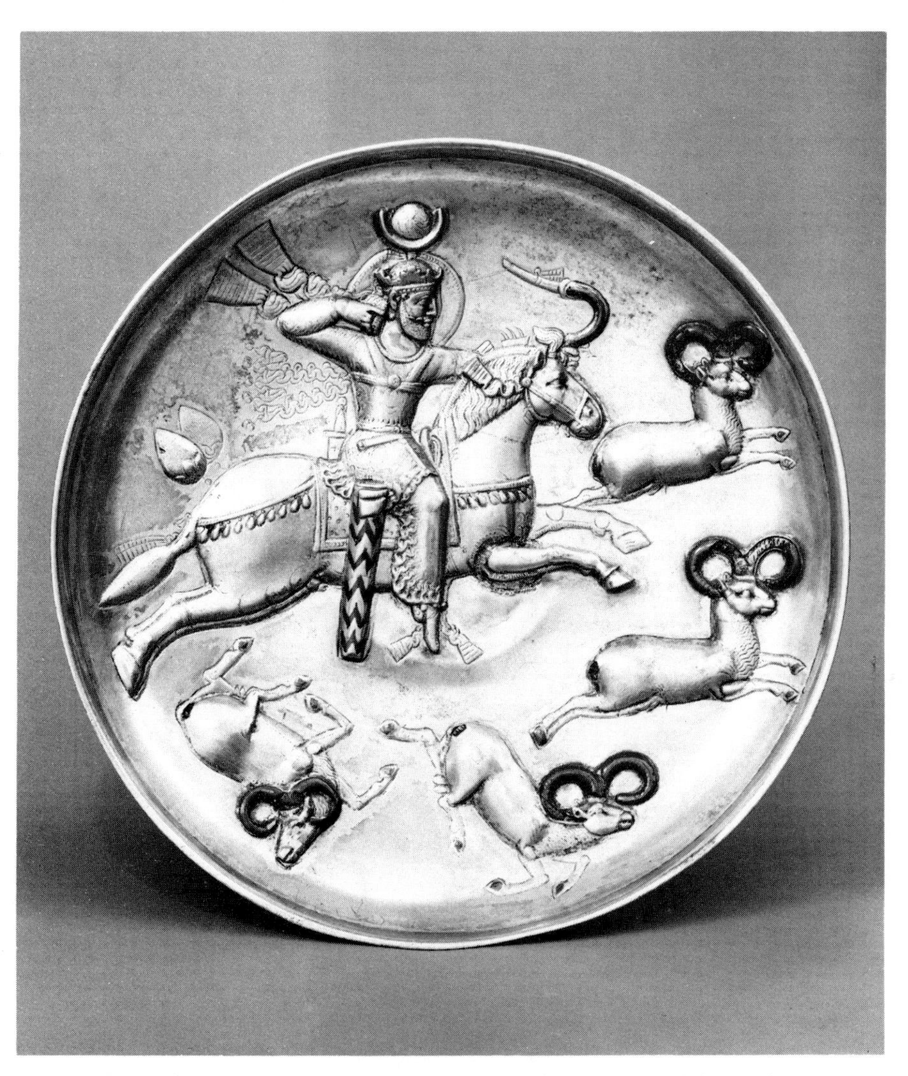

5. Persian hunting scene: King Peroz I (A.D. 459–84) shooting ibex with bow and arrow. Silver cast, engraved, embossed, and inlaid with niello. Hunting as the sport of kings represents an extremely long tradition in Near Eastern art. Hunting contributed to the decline of wildlife in the entire region. *The Metropolitan Museum of Art, Fletcher Fund, 1934*

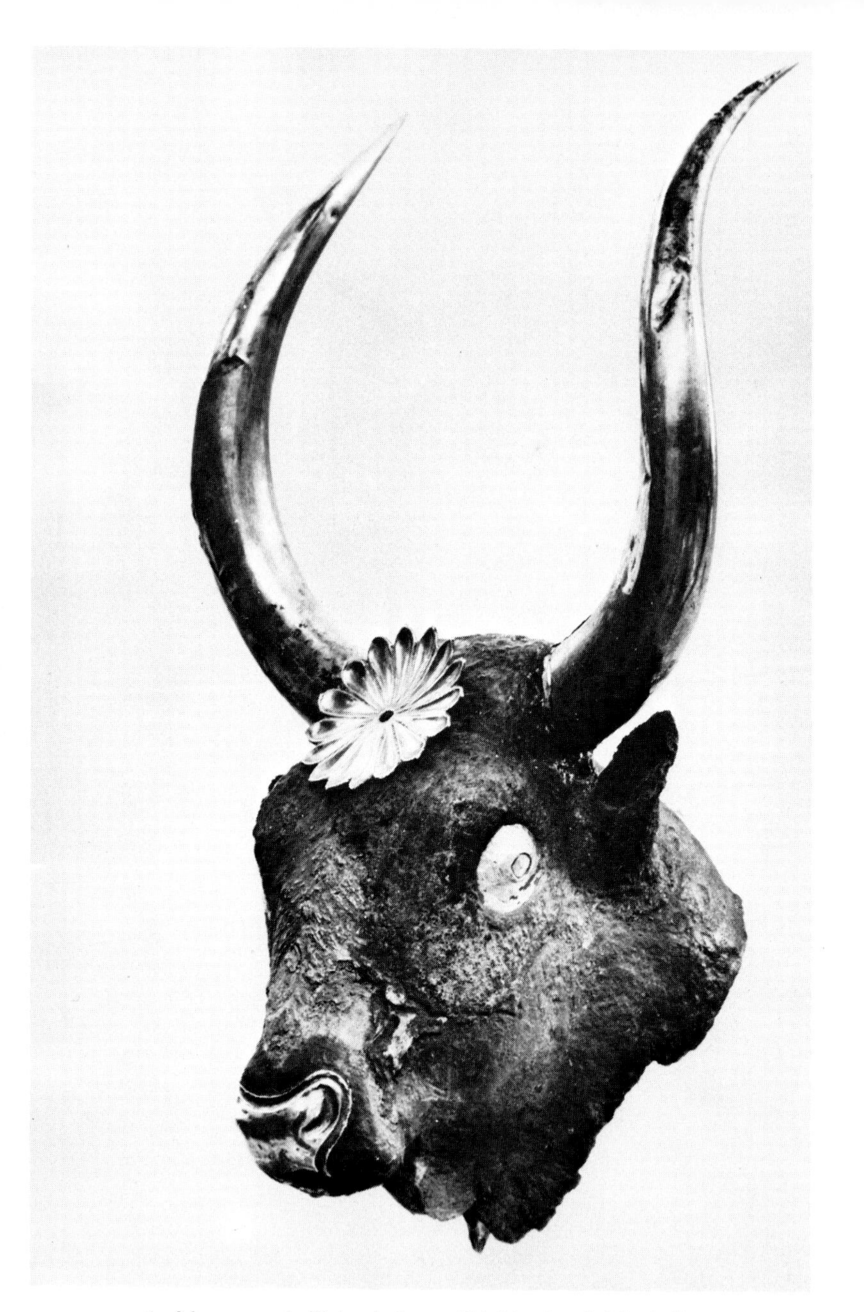

6. Mycenaean bull's-head rhyton (drinking horn). The bull was important in Neolithic and early Greek religion. The flower on the bull's head may be a fertility emblem, a sacrificial garland, or a reference to the sun, showing possible influence by the Egyptian Apis bull bearing the disc of the sun between its horns. *National Museum, Athens*

7. Attic amphora, black on red, early 5th century B.C. A representation of the olive harvest. Figure at right beats the branches of the tree with a pole to bring down the fruit, while figure at left puts the already-pressed oil into a bottle. *Vatican Museum, Rome*

8. Athena flying her owl, Greek statuette, bronze, 5th century B.C. The bird as an attribute of Athena ties the goddess to earlier nature deities of Crete. Most Greek gods and goddesses are intimately associated with the natural world. *The Metropolitan Museum of Art, Harris Brisbane Dick Fund, 1950*

9. Greek girl with pigeons. Grave relief, marble, from island of Paros, c. 450 B.C. It is clear that the birds are pets; this offers evidence of a positive, affectionate attitude toward animate nature during the Golden Age of Greece. *The Metropolitan Museum of Art, Fletcher Fund, 1927*

10. Greek sculpture of a lion, second half of 4th century B.C., Pentelic marble. Lions were still known and hunted in Greece at this time, and the artist may also have observed animals brought back from Egypt or Asia by expeditions associated with Alexander the Great at about the same time. The sculptor's technical skill in naturalistic portrayal of his subject is evident. *Nelson Gallery, Atkins Museum of Fine Arts, Kansas City, Mo.*

11. Fishing boats in the harbor of the island of Skopelos. The nets and the form of the boats with removable masts are virtually identical to those used in ancient times.

12. Model of Miletus, showing the city planning attributed to Hippodamus of Miletus. Hippodamian straight streets forming rectangular blocks were the forerunners of those in most Roman and American cities. *State Museums of Berlin, East Germany*

13. Herd of goats, mosaic from Hadrian's villa, 2d century A.D. The destructive grazing of goats was one of the most powerful forces of environmental deterioration in the ancient world, causing deforestation and erosion. These goats are apparently grazing in a sacred grove marked by a few trees and the statue of a goddess. *Vatican Museum, Rome*

14. Matera, Basilicata, southern Italy. Pronounced example of erosion and overgrazing; note horizontal marks in background, caused by grazing animals.
Publifoto, courtesy Italian Cultural Institute, New York, N.Y.

15. Cosenza, Calabria, southern Italy. The town is situated on a rocky ridge to avoid encroaching on arable land. Note erosion and use of terracing. *Italian Cultural Institute, New York, N.Y.*

16. Detail of Serpentius mosiac, Borghese Gallery, showing a *venatio,* or hunt staged in an arena. Serpentius has speared a stag; a bull approaches. *German Archaeological Institute, Rome*

Gardens and parks represent planning on a small scale to create a pleasing environment, a retreat from the hot sun, dryness, and dust, a place cared for and arranged by people to improve their style of life. Although the courtyard of a typical Greek house was adorned with pots of growing herbs and flowers, it could hardly be called a garden. Near country houses and residences of the very rich in the city, gardens were planted with trees, flowers, and shrubs and supplied with flowing water. A people whose myths included the Garden of the Hesperides certainly delighted in these adaptations of nature. Greek writers indicate that private gardens were numerous. In fifth-century Athens, Kimon threw his own garden open to the public, and parks were common from that time on. Marketplaces were commonly planted with shade trees. Epicurus met with his disciples in a private garden, and his school was thenceforth known as the Garden School. In later Greek houses, gardens were often planted in the peristyle, thus becoming an integral part of the house.

Greek cities needed a dependable supply of pure water, and in fact most of them were founded near springs. But as the population grew, many of these springs became inadequate and some dried up. Wells, dug in great numbers inside the cities, lowered the water table and were subject to pollution and eventual exhaustion. Many cities, therefore, constructed aqueducts, generally in the form of covered canals, to supply the public fountains. The people of Samos excavated a remarkable tunnel to bring water under a mountain to the city, and the aqueduct of Pisistratus at Athens runs underground for much of the way. To assure the purity of the water coming into the city, the aqueducts were provided with large settling basins where dirt and sediment would sink to the bottom. Water was not supplied to private homes, and except for rich households who were lucky enough to have their own wells, all city dwellers had to use water from public springs and fountains, which were often within ornate buildings. One in Athens was called Enneakrounos, or "Fountain with Nine Spouts." The city governments carefully supervised the water supply, and major users generally had to enter into contracts. Water theft, a fairly common offense, was punished by fines. The famous Themistocles once served as water commissioner at Athens and collected so many fines that he was able to erect a statue of the water bearer for the city. Pollution of the city water supply could merit the death penalty. Provision for sewage disposal was made in Greek cities. Laws at Athens and elsewhere required that waste matter be carried outside the walls for a certain distance before it was dumped. Drainage sewers were often covered channels, and any excess water in the city supply might be used to flush them out. The main sewer at Athens drained into a reservoir outside the Dipylon gate, and from there the waste water was conducted through a series of canals to enrich the plains near the city, so that Athens provided fertilizer for her own fields.

The existence of air pollution in Greek cities is undeniable. Countless cooking fires, charcoal fires to heat rooms, the smoke from metal working and pottery firing, not to mention the ubiquitous dust, meant that a city could be seen a long way off because of its smoke. Smoke to Greek authors is usually the first sign of human habitation. Temperature inversions, common in Greece today, are not a result of the industrial revolution; they held smoke and dust in suspension over ancient cities as over modern ones. The difference is in the chemical nature of the pollutants and their amount.

The Hellenistic Age was a period of rapid urbanization, as Alexander and his successors founded innumerable Greek cities in the eastern and southern kingdoms, even as far away as Afghanistan and India. The great new capitals such as Alexandria, Antioch, and Pergamum were planned with wide boulevards lined with colonnades, statuary, and trees. Gardens and parks, which added to the spacious feeling of the public areas, and larger buildings, such as palaces and temples, made these centers far more grandiose than the older city-states of Greece. Alexandria was laid out by Deinocrates, an architect who used the rectilinear pattern favored by Hippodamus but modified it by adding wider main avenues and canals, selecting dominating positions for major structures, and integrating the walls into his total plan. Most of these works, like the crescent of great buildings at Pergamum, were intended to impress the beholder with the might of the kings who commissioned them. Such lavish building created vast demands for stone, metal, and timber. So large and urban were the Hellenistic capitals that it is not surprising that their writers developed a literature of nostalgia for the unspoiled countryside.

Trade between the Greek cities was mainly by sea rather than land, since pirates and winter storms were somewhat less forbidding than mountains, bandits, and the customs and toll stations of countless independent cities. The Greeks constructed several large projects to make navigation easier. Canals were cut through narrow isthmuses in spite of the religious feeling against this alteration of the natural arrangement of land and sea. Where the isthmus was too wide and rocky for a canal, at Corinth, a causeway was built for ships to be rolled across between the Saronic Gulf and the Gulf of Corinth. Many great temples crown promontories like Cape Sounion; these served as navigation marks for Greek sailors, and were built partly for that purpose. The Pharos of Alexandria, a tall lighthouse, was erected on the long, low coast of Egypt, which lacks the bold capes of Greece. At times, signal fires were used by the Greeks to send news rapidly over long distances.

Greek roads were maintained by the cities, and were generally poor until the Hellenistic kings, inspired by the example of the Persians, began to improve them to provide land transportation across their wide dominions. Bridges spanned the rivers and such channels as the narrow Euripus Strait between the mainland and the island of Euboea.

The Greeks admired huge structures. It was they who compiled the list of Seven Wonders of the World, all distinguished by their huge size, and the majority of them built by Greeks. Strabo remarked in later times that the size and permanence of such works makes them part of the landscape itself, altering it and sometimes enhancing it.

Finally, military actions affected the natural environment. Wars were common in Greece, and were waged against the homes, crops, and resources of the enemy as well as his armies and navies. Homer spoke of warriors polluting the earth and the rivers with dead bodies and blood, and while his horror was at religious rather than environmental pollution, a battlefield after the conflict can hardly have been healthy. Vultures, jackals, and flies swarmed to such places, and both sides tried to bury their dead as soon as possible to avoid desecration. Theophrastus remarked that when an army had marched over a field of growing plants, hardly anything remained visible, and crops growing there the next year would be stunted. Of course, the actual battlefield, fertilized by the blood of the slain, might bear an unusually good crop the next year. Deliberate environmental warfare like that waged by the Spartans in Attica, where they devastated farms and fields and chopped down the olive orchards, was not rare in Greece. Invading armies often set forests on fire; this is one of the acts of impiety Herodotus attributes to the Persians. At the Battle of Pylos the burning of all the vegetation on an island made it impossible for the Spartans to move without being observed by the Athenians. War created demands for raw materials, wasted skilled farmers, reduced people to slavery, and destroyed whole cities. In addition to its obvious destruction of human values, its effect upon the natural environment was widespread and devastating.

9

Roman Attitudes toward Nature

The Romans loved their native soil; they were characteristically oriented toward the land. Earth, they firmly believed, was mother of all, and they called her *mater terra.* Never great seafarers, the Romans had found in their earlier history that the rich land of Italy met all their essential needs, and their attitudes toward nature reflect this. The earth may have provided for other nations, but this did not interest the Romans very much. They were confident that the earth of Rome and Italy was the best, being fertile, centrally located, and blessed with an equable climate. More than one Roman author claimed that Rome's conquest of the Mediterranean world was due to these very factors.

The Roman mind was marked by its practicality, and Roman attitudes toward nature were distinctly utilitarian. The Romans generally believed that the world is here for human use, and they proceeded in a very pragmatic way to find uses for its components. This dominant Roman characteristic is reflected in religion, literature, philosophy, and science. They sought knowledge for its own sake very rarely, always seeming to demand a practical application for intellectual endeavor. Closely allied to practicality was a profound desire for order; this was demonstrated in their view of nature as everywhere else. No other people in ancient times imposed a structure upon their natural environment so rigorous and artificial as Roman centuriation, a geometric system of dividing the land into a checkerboard of squares of equal size, which they used from the Po Valley in the north to Tunisia in the south. The squares are still clearly visible in aerial photographs.

The Roman attitude toward nature reflected Roman religion, which from early times contained a dominant strain of animism, the feeling that spiritual powers manifest themselves in the natural environment. The early Romans were an agrarian people with strong ties to nature, and their religion, being intensely conservative, preserved agricultural and animistic elements even among the city-dwelling populations of later times.

Roman religion always had a strong sense of locality. Certain places seemed to the Romans to be endowed by nature with supernatural powers, or numina. Temples or shrines came to be located in these places, but the divine power was felt to be inherent in the place itself, as part of its natural functions, before any special dedication by men for religious purposes. Ovid spoke of such a place in the *Fasti:* "Under the Aventine (hill) there lay a grove black with the shade of oaks; at sight of it, you could say, 'There is a spirit here!' "[1] Not only groves, but individual trees, rocks, springs, lakes, and rivers were regarded in this way. The area of volcanic activity near Naples, with its sulfurous caverns, steam vents, hot springs, and shaking ground, was thought to be particularly numinous, and famous oracles were located there.

Natural objects which provoked wonder and respect among the Romans included springs of water like the Fons Bandusiae, subject of one of the odes of Horace, and trees like the *arbor felix,* under which custom required that the hair shorn from the priest of Jupiter be placed. Groves of trees, as noted above, were regarded as the dwelling places of spirits and, as in Greece, were somewhat protected from damage. For example, horses were excluded from the grove of Diana at Aricia. But Roman practicality asserted itself here, and Cato the Elder records a prayer to be used with the sacrifice of a pig before cutting down trees in a sacred grove, to smooth the ruffled feelings of the god or goddess, whichever it might be, who lived there. A prayer like this one would, of course, lessen the protection extended by religion to a sacred grove. In similar fashion, a sacred stone could be carted away to a more convenient spot, carrying its numen with it.

Roman religion was not without reverence for wild places, forests, and mountains, but it was predominantly agricultural, reflecting the early observances of farm families close to the land who depended on the cycles and changes of nature for survival. The Roman calendar, therefore, was based upon the agricultural year, and its complicated series of religious festivals was dictated by the round of activities on the ancestral farm, from the hanging of the plow on the boundary marker in the Compitalia in January to the festivals of Saturn and Bona Dea, both deities of the soil, in December. When Virgil addressed his native land of Italy in poetry, he did so with the images of the old agricultural religion:

> Hail, great mother of harvests! O land
> of Saturn, hail! Mother of Men![2]

Romans always worshiped the great gods of Indo-European polytheism; it can be demonstrated through linguistics that names of gods like Jupiter were brought to Italy with the ancestral Latin language itself. These gods were associated with the natural environment, particularly with the aspects affecting agriculture. For example, Jupiter was regarded as the bringer of rain. But to the gods brought to Italy by the Romans' linguistic forbears were added many more of local origin. In fact, Roman gods and goddesses became innumerable, and their hierarchy ran from great deities like Diana, goddess of the woodlands and wild creatures, to local spirits of springs like Juturna. The Romans possessed gods of the farmhouse and storehouse—the penates—and of the fields—the lares. Anything that could be given a name seems to have had a deity in charge of it. A god or goddess looked after every major agricultural crop: Ceres was goddess of the growing grain, Liber god of wine, Robigus would protect the crops from diseases, and so forth. Beyond this, every major and minor activity of the farm had a deity who could be propitiated for its success, such as Vervactor for the first plowing, Repacator for the second plowing, Imporcitor for harrowing, Insitor for sowing, and so on, including even Sterculius for manuring. Long lists of such deities have been preserved. On the margins lurked the wild gods of mountains and forests: Silvanus, the fauns, and others.

Roman religion was not much concerned with careful definition of the nature and role of the gods, or with matters of ethics. It was concerned with ritual, and the ritual was designed for a specific practical end, that is, to control the natural environment through enlisting the cooperation of the unseen powers which control it. When a Roman addressed one of his gods, he might use the expression, *do ut des,* "I give so you will give." The prayer and sacrifice were offered in order to induce the god to grant a desired result, not simply to honor the god, and certainly not out of love. An offering might be made as a return for a favor already received, or because a previous promise had been made to give the god something if he granted what was desired. Presumably, if the god did not answer such a request, he would get nothing. Efforts to control nature through ritual were highly persistent among the Romans, for whenever they did not succeed, the failure could be blamed on some fault in the ceremony. It was believed that if the person repeating the ritual made any mistakes in words or actions, the ritual would lose its effect and therefore would have to be repeated. Persons whose presence might pollute a sacrifice, such as murderers and most foreigners, had to be excluded. Those present had to observe ceremonial cleanliness and silence.

The Romans sacrificed to the gods only such things as contained the principle of life, and with very few exceptions, only the animals and plants of agriculture and their products. Cakes, fruit, wine, cheese, milk, and other products could be offered, as well as the bloodier sacrifices of pigs, sheep, and oxen, often one of each, representative of the farmer's herds. In addition to sacrifice and prayer, the Romans practiced a multitude of ritual actions intended to control the environment and drive away evil. Many of these actions amounted to magic, such as circumambulation, which drew a sacred circle around the fields and excluded baneful influences.

Another Roman practice that should be mentioned is divination, which rests on the belief that the natural environment reveals the will of the supernatural powers to people. In particular, the flight and calls of birds, the actions of animals, especially wild animals, and the occurrence of lightning and thunder were observed and interpreted by men skilled in these matters. The internal organs of sacrificial animals, particularly the liver, were examined for unusual markings and shapes. To the Romans as to other ancient peoples, the world was full of gods, and human beings were wise to tread carefully. The feeling that the natural environment is the result of divine arrangements and ought not to be altered unduly or in massive ways lurked in the minds of the common people and produced a conservative attitude in this respect, although it was not strong enough to prevent the execution of many of the spectacular works of engineering planned by the Romans. Tacitus does say that a planned diversion of the waters of the Tiber was halted by the expression of just such feelings, but this was a rare exception.

In religion, as in much else, the Romans felt the deep and pervading influence of the Greeks. At an early time, the Roman gods were identified with Greek gods who seemed roughly similar, and myths of Greek gods were freely ascribed to their

Roman counterparts. As the centuries went by, Roman and Greek ideas became inextricably mixed, so that much of what we said in chapter 7 concerning Greek attitudes would be true to some extent of Roman attitudes as well, in the period after the Roman Empire had been extended into the Greek world.

Religious influences also reached Rome from Asia Minor, Egypt, and Persia. Astrology in particular, with its implicit theories of environmental determinism, became exceptionally popular in Rome and affected Roman thought about the workings of the natural world. Astrology teaches that human life is determined by the changing positions of the stars and planets. Various zones of the earth are considered to be under the influences of different constellations of the zodiac and planets, which accounts for the differences in peoples and animals found there. Such a system of belief leaves little room for the idea that human beings are capable of producing changes in the natural environment, and thus it tends to make the study of human interaction with nature—or at least that part of nature which occurs on earth—pointless.

Roman literature contains countless celebrations of the beauty of nature, in both its wild and domesticated forms. In fact, this is a dominant and recurring theme. As Pliny the Younger remarked, "There is nothing that gives either you or me as much pleasure as the works of nature."[3] It must be noted that in this respect the Romans were following a literary tradition of nature description which they had inherited from the Greeks of Alexandria, so that some of their expressions may reflect convention rather than feelings original to the author, but the Romans had their own traditions of agricultural writing going back at least to Cato the Elder. So highly did the Romans regard good writing in this field that the Senate, in ordering the total destruction of Carthage at Cato's urging, made an exception of the book on agriculture by Mago, a Carthaginian. This was translated from Phoenician into Latin and, although too late to affect Roman practice during its best yeoman phase in the days of Cincinnatus, it did influence the classic Roman writers on agriculture: Cato, Columella, Varro, and Virgil. Literary tradition alone could not explain the richness and variety of nature imagery in Roman writing; most of the authors had really observed nature, and genuinely took delight in the beauty of their natural environment.

To a great extent, Roman writers probably describe scenes which were familiar to them since childhood. "If we have dwelt some time amid mountains and forests we take delight in them,"[4] observed Cicero. Many of the authors had moved to Rome from towns in the Italian countryside, which helps to explain the depth of their feeling for rural settings. A passage from Virgil typifies this kind of landscape description:

> For now the farmhouse gables are smoking in the distance,
> and larger shadows fall on the lofty mountains.[5]

In the great age of Latin literature, Rome was highly urbanized, and it became common for Roman writers to draw a contrast between city and country life, stressing the advantages of the latter and vicious effects of the former. It was a genuine conviction among the Romans that rural life was morally and physically superior to urban life. Horace wrote that he disliked going to the city of Rome and wanted to retire to a lovely spot in the country, and Juvenal's descriptions of the city are positively horrifying in their verisimilitude. In the *Georgics,* Virgil describes with evident delight the humble pleasures of farming, the round of the seasons, and the duties of the husbandman. There is no such praise accorded to the amenities of the capital city.

The glories of wild nature are also sung by the Roman writers. "May I love the streams and the forests!"[6] Virgil entreats. Elsewhere he describes streams tumbling down rocky canyons, and he loves to name mountain peaks. Although mountains are described in terms of gloom and dread by some Romans, others enjoyed studying them and climbed them to obtain wide views. Lucretius speaks of mountain climbing as a personal experience, Seneca asked Lucilius to climb Etna to make observations for him, and it is recorded that the emperor Hadrian himself climbed Mount Etna and Mount Casius to see the sunrise, and at least one other peak as well. Curiosity and the desire for aesthetic pleasure seem to have been the primary motives for these exploits.

More evidence of Roman enjoyment of nature comes from painting and mosaic art, where artists (probably following Greek models which have been lost) produced outdoor scenes with startling naturalism. The walls and floors of Pompeii and the palaces of the emperors are covered with artists' renditions of trees, mountains, seascapes, lifelike birds, mammals, and creatures of the sea. Sometimes objects from nature are treated in a highly stylized or grotesque manner, but often the attempt has been made to fool the eye into believing that it is seeing the object itself. Here nature is treated as a background to human life.

In philosophy, the Romans were followers of the Greeks, being influenced by the schools of Plato, Aristotle, and Epicurus, but most of all by the Stoic school founded by Zeno and developed by Chrysippus and others.

The Stoics taught that the natural world represents the design of divine reason. Indeed, according to Stoic pantheism, God not only designed the world but pervades every part of it. He is the soul of the universe. In such a world, everything has its place and purpose. As Cicero says in his explanation of Stoic philosophy:

> Unless obstructed by some force, nature progresses on a certain path of her own to her goal of full development . . . in the world of nature as a whole there must be a process towards completeness and perfection. . . . There

> can be nothing that can frustrate nature as a whole, since she embraces and contains within herself all modes of being. . . . Since she is of such a character as to be superior to all things and incapable of frustration by any, it follows of necessity that the world is an intelligent being, and indeed also a wise being.[7]

The world is directed by its own principle of reason, and everything in it has a rational purpose. Plants exist for the sake of animals, animals for the sake of mankind, and mankind exists to contemplate and imitate the perfection of the world.

Not all Roman philosophers accepted this view of nature. The Epicureans, represented by the poet Lucretius, held that the world is the result not of design but of chance, a configuration of atoms in the void which is doomed to change and decay. Animals and plants, some Epicureans held, could not have been created for human use, because so many people are fools, and there is not enough human intelligence in the world to make creation worthwhile. Not many Romans seem to have accepted the Epicurean view that the natural world is driven by chance, essentially mechanistic, and purposeless except as it simply follows its own blind path. But many did accept the idea that the world is growing old and decaying, that the soil is less fertile than it was, and that history is a process of decline. Here Epicurean philosophy seemed to give support to the mythical doctrine that the world has been getting worse ever since an ancient Golden Age, described by Hesiod among the Greeks, and in Roman tradition represented by the fabulous Reign of Saturn, when a fertile earth had borne crops by herself.

The Romans had an opportunity to observe a fairly large segment of the earth. In conquering and ruling the Mediterranean Basin and lands beyond it as different as Britain and Mesopotamia, they observed various climates and the races of mankind who lived within them. The conclusion made by many of their geographers and philosophers was that the environment shapes the people who live within it, a conclusion which was probably suggested to them by the writings of Hippocrates and other Greeks. This doctrine of geographical and climatic determinism has had a long and tenacious life in the history of the study of civilizations. To the Romans, it served as an explanation and vindication of their rule over other peoples, for as Rome was obviously in the center of the world and possessed the best of all possible climates, her citizens could not help but be superior to others, and inevitably gain ascendancy over them. Aristotle had made a similar claim in behalf of the Greeks. Environmental determinism does not explain all the observed differences between peoples, of course. Some groups who live very close to one another, in virtually identical climates, have quite different characters and customs. Polybius pointed out that neighboring tribes in Arcadia had divergent temperaments; he ascribed the difference to varying cultural traits. Custom and culture, he maintained, not just the environment, determine how people will behave, and many Roman writers agreed with him. To these factors, others attempted to add the astrological determinism which has been described above.

Counterposed to the idea that the natural environment determines the expression of human nature is the idea that mankind can change the environment. The Roman philosophers gave expression to the latter belief as well as the former; Cicero, in praising the cleverness of human hands, describes the many things that they can do for us, including agriculture, domestication of animals, building, mining, forestry, navigation, and hydrology, and concludes with this insight: "Finally, by means of our hands we endeavor to create as it were a second world within the world of nature."[8] The Romans, as much as or more than any other people of ancient times, exerted themselves to create this "second nature" of human devising, and they could see the impressive results of their own labor everywhere.

Cicero's statement is another illustration of the utilitarian and manipulative bias of Roman attitudes.

Stoic philosophy saw human ability to change the environment as resulting from mankind's participation in the rational, creative life of the natural world itself. The rational design of the world includes provision for human activities; as Seneca noted, metals are hidden in the earth, but mankind possesses the ability to discover them. In this view, man is the natural caretaker of the earth, and its creatures are commended to his care. Rational human efforts make the earth more beautiful, which is to say, more serviceable for human purposes, and, in a typically Roman view, beauty and utility are synonymous. Mankind improves animals and plants through domestication. In the same way, the extension of civilization is seen as making up a deficiency that exists in the wilderness, which is seen as a "haunt of beasts" or a "barren waste." Epicureans did not share the Stoic faith in a designed world; they saw human efforts rather as a struggle against the tendency of the world to decay. But the resulting attitude is much the same; wild beasts are seen as menaces, and their success in the struggle for existence, according to Lucretius, results from cunning and courage, or domestication and the protection of mankind.

On the other hand, Cicero could maintain that "the products of nature are better than those of art,"[9] and many Romans were aware that human activities often produce results which are neither beautiful nor useful. Pliny the Elder complained that people abuse their mother, the earth. The doctrine that the earth is growing old and decaying was attacked by the wise agriculturist, Columella, who said that the blame for nature's infertility lies not in some supposed senescence or "the fury of the elements," but in poor husbandry.[10] A good farmer knows how to restore the soil to fertility, but those who misuse the earth should not be surprised when the result is diminishing crops and sterility. Roman attitudes could encourage wise use of the earth's resources, with an eye to sustained returns in the future.

One attitude toward the natural environment which is of the greatest importance to ecology is curiosity, the desire to understand how nature works. Romans were interested in nature and curious about it, but they were makers of practical observations, seldom true scientists. Pliny the Elder and others made endless collections of unusual facts and fictions about nature, but failed to verify their reports by observation.

Much of Roman science was inherited from the Greeks, and in some fields, as in medicine, it remained in Greek hands during the period of Roman ascendancy. In agriculture, architecture, and mechanical technology, the Romans made some advances which bear upon ecology, and these will be discussed in the following chapters.

10

The Impact of Roman Civilization on the Natural Environment

The changes made by the Romans in the natural environment were striking and far-reaching, because Rome exceeded all the earlier civilizations of the Mediterranean in technical skill and ability to organize, and because the area which came under Roman domination and influence constituted the largest empire in history up to that time. Avaricious and practical, with an activistic willingness to exploit others and the world around them for their own profit, the Romans left their mark on the earth. It is a mark traceable not only in their native Italy, but particularly in the newer lands of the west and north, where technically less-developed societies gave way before their arms and their business schemes. The Roman mark is present also to a surprising degree in the east and south, where older civilizations had already altered and depleted the natural environment to a much greater extent. Roman demands on nature were felt even beyond the imperial frontiers, where trade, military forays, and the introduction of new technology carried Roman influence.

The loss of forests was the most widespread and noticeable change made in the natural environment by Roman activity. Deforestation proceeded more slowly in the northern and western areas, where rainfall was greater, than in the eastern

92348

Mediterranean Basin. Two generations after Plato had lamented the wasted forest of Attica, Theophrastus could still report that Italy and Sicily were well forested, Latium in particular. He also relates the charming story that when the first settlers came to Corsica, they found tall trees growing so thickly and so near the shores that the masts of their ships were broken by great branches which extended out over the water. Roman writers, too, indicate that Italy had extensive forests, some of them so thick and dangerous as to be almost impassable, such as the Ciminian Forest in Etruria. In earlier times, religious feeling may have inhibited the cutting of trees, but Cato recorded a prayer to be used before clearing trees even from a sacred grove, beginning with the formula, "Whether you are god or goddess, to whom this grove is consecrated. . . ."[1] Forests were the major source of fuel for the Romans, who either burned the wood or reduced it to charcoal. Much timber was used in construction, although evidence from architecture indicates that wood was gradually replaced by brick and stone in Roman buildings, probably due to the shortage of lumber and the expense of importing it. Shipbuilding used large amounts of forest products, including wood and pitch, particularly during and after the wars against Carthage, when large Roman navies were built. More wood was required for military operations on land, especially for siege engines, and for transportation in general.

All these demands resulted in the disappearance of forests from the settled areas and accessible coasts. Location away from cities and ports, in rough terrain, tended to preserve forests by increasing the cost of transportation. Forests in the lowlands were cleared for cultivation; Lucretius remarked that people were forcing "the forests to recede daily higher up the mountainside and yield the ground below to agriculture."[2] Even the mountain forests were not safe, however. Herders set them on fire to improve the grazing. Logs from the mountains were floated down the Tiber River to Rome, and similarly down rivers in other parts of the Roman Empire. The best shipbuilding ports were located at the mouths of rivers draining forested mountain slopes, where the logs could be floated down. Masts had to be made from tall, straight, strong trees, and these were not easily found except in mountainous regions. Pines were imported to Rome from the Black Sea. Tiberius Caesar had larch logs, one 120 feet long, brought from the Raetian Alps.

The Romans made some efforts to develop plantations of trees grown for wood, as in woodlots on farms, and sometimes used irrigation for this purpose. But the major proportion of wood used was imported from various parts of the Roman Empire in a widespread and important sea trade. Heavy lumber was transported chiefly from mountainous areas of heavier rainfall in the north, while specialty woods came even from the east and south, such as cedar from Lebanon and "citrus" from North Africa.

Because the lumber trade was very important to the military, the Roman government promoted and regulated it. State forests were rented out to private exploiters. Cutting off shipments of timber could be a weapon against actual or potential enemies, and was used against Rhodes in the first half of the second century B.C.

The deforestation of much of Italy and the Roman provinces was unfortunately permanent. The porous limestone soils common in the region do not recover well from deforestation. After the removal of the natural cover, erosion proceeded rapidly in the short but torrential Mediterranean rainstorms, and goats usually ate such small trees as began to grow. There were frequent fires set deliberately by herdsmen, farmers, and soldiers.

Among the results of this permanent deforestation were flooding and greatly accelerated erosion. Without forest cover, the hillsides could not retain the water that fell on them, and flash floods descended on the plains and cities below. Rome suffered serious floods from time to time in the low-lying areas of the city, the Forum and the Circus Maximus in particular, and was impelled to develop an extensive underground drainage system to carry away the flood waters. The early Cloaca Maxima was intended primarily as a storm drain rather than a sewer, although it served both purposes, and was followed by other drains. The siltation of lowlands, lakes, and the Mediterranean seashore created new marshes and greatly altered the coastlines, in some cases pushing them many miles farther out to sea. This harmed many port cities. Paestum declined when her harbor silted in, and Ostia, at the mouth of the Tiber, was maintained only by the periodic construction of new harbor basins as the old ones were isolated or filled with debris brought down by the river. Much later on, Ravenna, the chief Roman port on the Adriatic coast, lost her access to the sea through a similar process. Erosion caused whole provinces—notably Syria and North Africa—to decline in ability to support their populations. The deposits of rock, sand, and mud washed down by Mediterranean streams after deforestation are up to thirty feet thick in many places. These materials were washed down from above, where the hillsides, deprived of their natural cover of vegetation, now lost their soil cover as well. Without the forests to hold back water and deliver it more slowly to the areas below, many springs and smaller streams dried up, as some writers at the time reported. During the rains, the water in some of the aqueducts was muddied. Marshlands created by deposits from erosion presented another danger in the form of malaria, which became a widespread disease in Rome early in the second century B.C. after its introduction, possibly from Greece. The lower country near the city of Rome supplies a notorious example of flooding, repeated development of swamps, and the endemic problem of malaria, due to upland erosion. The Romans periodically embarked on ambitious programs for the drainage of marshlands, destroying wildlife habitat as they did so, but many of the new alluvial deposits were too low-lying for efficient drainage, and were never completely reclaimed throughout the period of the Roman Empire. The drainage of the ill-famed Pontine Marshes was not permanently accomplished at any time before the twentieth century, because the basic problem of erosion and drainage was chronic.

The use of wild animals by the Romans constituted gross exploitation and may serve as evidence for their general treatment of nature. While their ancestors had lived by hunting, and some poorer people in outlying districts still did so, for the Romans in general hunting was a matter for sport or business. Roman mosaics illustrate the activities of the hunt, showing hunting dogs and falcons. Falconry with hawks and eagles was introduced from Persia. Hunting was a private sport in Rome, although certain privileges, such as hunting lions, were in time reserved for the emperor.

Romans with large estates usually had game parks set aside for themselves and their guests to hunt in, and constructed aviaries for game birds as well. They killed and ate many more species of birds than is common today. Not only the upper classes hunted. Roman peasants hunted smaller animals for food occasionally, and killed the predators who fed on their herds. The wolf, maternal symbol of the founding of Rome, was extirpated from the more thickly settled areas.

Rome demanded wild animals and their products from the provinces of the empire and beyond. Ivory from Africa and India was used in works of art including huge statues of ivory and gold, and was inlaid in furniture of every kind. Ivory writing tablets, desks, spoons, and other objects were popular among the Romans. Wild animal skins were used for clothing and furnishings, and feathers of ostriches and other birds served for decorations on military uniforms and elsewhere.

But the entertainment industry of Rome, which included the

display and killing of countless animals, was more wasteful and destructive of wildlife than any other feature of Roman culture. The shows put on in amphitheaters for the amusement of the people included animals who could do tricks and take part in pageants and plays to indulge Roman tastes, which ran heavily to sex and violence. Some of the rarer animals were merely shown as curiosities, but more often they were mutilated and killed. *Venationes,* or hunts, in which armed men, sometimes even the Praetorian cavalry, chased and killed animals, constituted a major part of the shows, and are often the subjects of Roman mosaics and paintings. Fights were staged between goaded and crazed animals—between a bull and a rhinoceros, for example. Unarmed or poorly armed people, usually condemned criminals including those guilty of the "crime" of belonging to an illegal religion such as Christianity, were exposed to starving animals such as lions, leopards, and bears. Special cages were constructed under the amphitheater, complete with elevators and ramps to bring the beasts up to the arena without endangering the attendants too much. Romans of every social level, from the emperor to the common people, attended the games and enjoyed a spectacle which can only be called sadistic, and Roman writers, with rare exceptions, describe the games with approval.

The variety of wild animals and the vast numbers killed must be of the greatest concern in this book. Only the larger animals were used in the arena, since they had to be visible to thousands of people at once. But all kinds of large mammals, reptiles, and birds were imported from the empire, Europe, Africa, and Asia as far as India and even Thailand. Elephants were first seen in Rome in 275 B.C., after Pyrrhus of Epirus brought them to Italy in his military campaign. Ninety years later, ostriches, leopards, and lions were seen, and in the last century of the Republic, hippopotamuses, crocodiles, and rhinoceroses made their entrance from Egypt, and Caesar sent a lynx from Gaul. Augustus displayed tigers from India, and Nero showed polar bears catching seals.

The numbers of animals killed are phenomenal, mounting into the hundreds in a single day. Augustus had 3,500 animals killed in 26 *venationes.* At the dedication of the Colosseum under Titus, 9,000 were destroyed in 100 days, and Trajan's conquest over Dacia was celebrated by the slaughter of 11,000 wild animals.

Since there were many amphitheaters throughout the Roman Empire, the demand for wild animals was enormous and was supported by an extensive, organized business for hunting, capture, and transportation. Many found employment in this enterprise, which was far from easy, since the beasts had to be kept in good condition in pits, nets, or cages from the time they were captured until they were delivered, as in the case of the City of Rome, to the menagerie or *vivarium* outside the Praenestine Gate. For the most part, it was a private business on which the Roman government levied an import tax of 2½ percent, and many of the animals went to private parks. Roman officials were deeply involved in the trade, however, and soldiers were sometimes used to round up the animals. Animals specifically captured for the emperor had their food and other needs requisitioned from the towns through which they passed —not a small charge for some towns on the usual routes, considering the size of the animals and their number. Only the emperor was permitted to own elephants, and had a special officer, the *procurator ad elephantos,* to keep the imperial herd at Ardea, but in Republican times at least one private citizen had made a custom of impressing his friends by riding an elephant when he went out to dinner. Emperors often kept impressive animals of various kinds in their palaces, and rich citizens sometimes had tame lions or other animals as pets. More ordinarily, they kept housecats, those terrors of mice and garden songbirds, which they spread from Egypt through Western

Europe. Granted all of this, it is surprising that the Romans did not follow the lead of the Hellenistic Greeks in establishing zoos. Certainly they directed little scientific study toward animals, although Galen and other physicians did take the opportunity of viewing the internal anatomy opened to their view by the mutilations of the arena.

The result of the Roman procurement of animals was the extinction of larger mammals, reptiles, and birds from the areas most accessible to the professional hunters and trappers. The Romans themselves boasted of this, pointing out that they were removing dangers to man and his agriculture. But they also exhausted the hunting grounds of North Africa, where the elephant, rhinoceros, and zebra became extinct. The hippopotamus was extirpated from the lower Nile, and lions from Thessaly, Asia Minor, and parts of Syria. Tigers disappeared from Hyrcania in northern Iran, the closest source to Rome. Of course, these creatures were hunted and killed by other people and for other reasons too, but the Roman demand for animals effectively aided in making large wild animals rare or extinct in the entire Mediterranean Basin.

Fish were an important part of the Roman diet. Fishing boats used nets, and towers were built to sight the schools of tunny and other fish. They were caught locally, or salted and imported from as far away as Egypt, the Black Sea, and the Atlantic coast of Spain. Landowners developed both fresh and salt ponds to raise fish, lampreys, and oysters for their own tables and for sale. Those lucky enough to own waterfront property extended their fish enclosures and traps out into the sea. Some of the income of the Roman government came from the rental of fishing rights on coasts, lakes, and rivers. The extent of depletion suffered by Roman fisheries is hard to determine, but they did import from great distances and resort to artificial propagation. Other products of the sea were sponges and the murex shellfish, which provided an expensive dye.

Mining was an extensive means of exploiting the natural environment in Rome. Having learned metalworking and mining from the skilled Etruscans, the Romans then conquered the Etruscan homeland and began to work the mines, extract the metal in smelters, and manufacture articles of metal. Some mines in Etruria produced thousands of tons of ore every year. As the Roman Empire grew, mines were established in every part of it, but particularly in Spain, Gaul, Britain, Macedonia, Dacia, and the Danubian provinces, and an important import and export trade in metals was developed with lands as far distant as India. The Romans took over the older mines in the eastern Mediterranean. Some of these had already been exhausted, but in the famous Athenian mines at Laurium the Romans were able to use better smelting techniques to rework the ancient slag heaps for silver and lead. This is a good example of early recycling; modern methods, in turn, have made possible a second reworking of the Roman slag.

The Romans used placer mining and hydraulic mining whenever possible, and open-pit mining in preference to underground mining. Their techniques were inadequate to work the poorer grades of ore. Thus, hardly any workable bodies of high-grade ore were left on the surface within the area exploited by the Romans. The labor in Roman mines became harder and greater technical skill was required as the richer and more easily obtained ore was exhausted, so that miners were forced deeper underground. Tunnels reinforced with timber or stone supports ran for miles and reached depths of about five hundred feet. Problems of drainage, air supply, and cost made greater depths inaccessible.

Miners' tools changed little through the centuries. The familiar metal pick and shovel, hammer and wedge were used. Rock was loosened either by setting fires against it and pouring water on it to cool it rapidly, or by the use of battering rams. Bodies of ore were undermined and deliberately collapsed by removing the supports. Waterwheels and the Archimedean screw pump as well as simple hand bailing were used to lift water out of the mines, but in many places streams were directed into the mines to wash the ore out.

Working conditions in Roman mines, as in ancient mines generally, were atrocious. Skilled technicians, free or slave, directed thousands of slaves and condemned criminals, including men, women, and children. The tunnels, quite narrow and often steep, were rarely high enough to permit any posture except crawling on all fours. Dragging heavy sacks of ore, the laborers were often beaten when they paused to rest. Cave-ins were common, and the air was impure. Vitruvius speaks of deaths underground from noxious gases, and recommends the use of a lighted lamp to test the purity of the air, hardly a good method if the gases should happen to be explosive. It is understandable that the miners rose in revolt several times, and the decision of the Roman Senate in the second century B.C. forbidding mining in peninsular Italy was probably motivated by a desire to remove this source of danger rather than any awareness of a need to conserve the resource. Conditions in the mines were only slightly alleviated by the social reforms of the more enlightened emperors of the second century A.D., when baths were provided for the miners, among other improvements.

Mining was of great concern to the Roman state, especially where it involved the supply of the monetary metals. The mines were generally controlled by the state but contracted out to large private companies who provided the labor and paid high and carefully regulated rents. The Roman Republic maintained a monopoly on salt mining and controlled all mining to some extent, taking political and foreign policy considerations into account. The emperor later maintained certain mines, particularly those producing gold and silver, as his own property, administered by an official called the *procurator patrimonii.*

Many Roman mines were exhausted and closed. This placed a

strain on the Roman economy and money system, depending as it did on coined gold and silver. The need for monetary metals became critical under the Antonine emperors, who conquered Dacia to exploit its mines. The tin mines of Spain gave out in the middle of the third century A.D., and the Romans had to develop the reserves of more distant Britain.

Roman mining had many environmental effects. It caused major erosion, removing hillsides and clogging streams. Poisonous substances such as lead, mercury, and arsenic were mined. Streams diverted through mines or polluted by seepage from mines must have carried poisons downstream and often into the fields, where their waters were used for irrigation. Mercury was smelted by a process which produced mercury vapor, and was used in gold refining and as an adhesive for gold leaf. Arsenic was used in pigments and medicines, and its poisonous properties were recognized.

Serious and widespread problems with lead poisoning in Rome have been pointed out by Dr. S. Colum Gilfillan,[3] who lays particular stress on the use of lead kitchen utensils and the insidiously cumulative and long-lasting effects of lead poisoning, which would not always be connected with the cause by the sufferers. Of course, some Romans did recognize the poisonous properties of lead. Lucretius and Vitruvius both described the ill effects suffered by lead miners, and Vitruvius went on to warn against the use of leaden pipes in aqueducts:

> We can take example by the workers in lead who have complexions affected by pallor. For when, in casting, the lead receives the current of air, the fumes from it occupy the members of the body, and burning them thereon, rob the limbs of the virtues of the blood. Therefore it seems that water should not be brought in lead pipes if we desire to have it wholesome.[4]

Lead pipes will contaminate water if the water is acidic. Certain bacteria often found in water systems will provide the necessary acidity and put the lead into solution. The Romans produced far more lead than any other ancient people before them and, through the smelting process, added measurably to the concentration of lead compounds in the earth's atmosphere. Strabo reported that lead-smelting furnaces were provided with chimneys "so that the gas from the ore may be carried high into the air, for it is heavy and deadly."[5] This is evidence of severe local air pollution problems in places near metal refining furnaces in Roman times.

Quarrying for the vast building projects of the Roman Empire—including the famous roads, walls, and military fortifications as well as public and private buildings as large as the Colosseum and the temples at Baalbek—was carried on in ways similar to mining and had similar effects. Most quarries were open excavations in hilly country, with all the disadvantages of strip-mining. The Romans sought not only the pure white marbles favored by the Greeks, but also brilliantly colored and patterned marbles, granites, and porphyry. In addition, they used vast quantities of clay for bricks, ceramic tile, and pottery. Limestone, sand, and plaster were required for mortar and concrete, two Roman advances in the construction industry. These desired materials were shipped throughout the Roman world and often used far from their point of origin. Roman shipwrecks containing columns and other architectural and sculptural masses have been found in the Mediterranean.

Agriculture was by far the most important, most widespread Roman economic activity, and—with forestry—had the greatest effect on the environment. Despite the considerable urban development of the Roman Empire, Rome always remained a basically agrarian society dependent upon the soil and its products. The majority of the people were employed in agriculture, which was also the most important sector for financial investment. The fortunes of Roman agriculture, therefore, affected all the other activities of the Romans, and would constitute an excellent indicator of the degree to which the Romans maintained or failed to maintain a healthy relationship to the natural environment.

Roman agriculture was divided into two major kinds of activity, stock raising and agriculture proper—that is, the work of the herdsman as distinct from the work of the farmer with sown and planted crops, orchards, and animals which can be raised locally. The raising of sheep, goats, and, to a great extent, cattle involved transhumance, the annual shift to moister pastures in the mountains during the long, dry Mediterranean summer. At lower elevations, pastures would need irrigation, but the scarce water was usually needed for other purposes. Overgrazing was a constant practice in the Roman homeland, as the entire Mediterranean. The usual mixed herd of sheep and goats could effectively remove the plant cover from the hillsides; the sheep would eat the grass and herbs, including the roots, and the goats would severely browse the shrubs and trees. The resultant erosion could be disastrous, especially when combined with deforestation or fires, and the shepherds often caused both, in order to open the hillsides for grazing. Grazing animals do return manure to the soil, but this is of genuine value only in conditions of well-regulated grazing. The manure of grazing animals was usually lost to the farms during the six summer months when the herds were in the mountains. Sometimes sheep were allowed to graze among the olive trees, but goats damaged both olives and grapevines, and laws limited the places where they could graze. Pigs were herded in the oak, beech, and chestnut forests, where they ate the acorns and whatever else they could find, or were raised on the farms. They provided the most common meat for the tables of the Romans, who generally ate more meat of all types than the Greeks did. Horses could be raised somewhat more easily in Italy than in Greece, but they remained a luxury. The rich middle class of Roman citizens was called "Equestrian." Horses were trained for war, and for racing in the Roman *circus.* Much of the stock had to be fed for at least part of the year, so some of the land was used to raise fodder. Alfalfa, which had been introduced to Greece from Persia in the fifth century B.C., was brought to Italy in the second century B.C. The spread of new plants, cultivated and wild, is an important human alteration of various ecosystems.

Early Roman agriculture was very similar to that already described for Greece. Farms were small, and the land used intensively and specifically for the purposes for which it was judged best suited. But as Rome expanded its territory, the landholdings of the richer Romans increased greatly in size and were converted generally to stock-raising ranches. Less acreage was used for grain, which was instead imported from provinces outside Italy, and much land which would have been useful for farming was devoted to grazing. This whole process was aided by the Punic Wars, in which Rome gained overseas territories from Carthage and at the same time lost hundreds of small farmers. The campaigns of Hannibal, in the Second Punic War, devastated and depopulated much of rural Italy. Rome, victorious in the war, allowed richer citizens to gain control of most of the emptied lands, legally or illegally. Some of these lands were developed into plantations worked by slave gangs, but most became livestock ranches, also run economically by a few trained slaves. Slaves were then plentifully provided by the large numbers of captives taken in the wars. By the succeeding century, the Roman population was restoring itself, but land was not available for these people because it had already been appropriated by the wealthy. Agrarian reform, led by the brothers Gracchi, attempted to distribute some of these preempted public lands among ordinary Roman citizens. Redistribution would have reestablished a reliable group of small farmers, subject to periodic military service, on the land and would incidentally have reconverted that land from pasture to intensive farming. The reforms were largely frustrated by the landowners, and the Gracchi brothers were assassinated, but in the first century B.C. much of the same effect was achieved in some areas by land grants to military veterans. Massive importations of grain to Rome from Sicily and North Africa, subsidized by the Republic and sold below cost to the common people, were begun by the Gracchi, and in this point their reform succeeded, undermining the economy of grain production in Italy and encouraging other uses of the land. Near cities, market gardens developed, raising perishable foodstuffs for local consumption. Under the Roman Peace of Augustus Caesar and the first few emperors who succeeded him, agriculture flourished in the vast sphere of activity furnished by the enlarged empire. Forest lands were cleared, often with the encouragement of the imperial government. Irrigation and drainage works were constructed in Italy and in all the provinces. Italy did not need to be self-sufficient in agriculture, because she could supply herself through imports from the various provinces. Wheat was now provided by Egypt, whose seldom-failing Nile floods made her the granary of the Mediterranean. Extensive olive groves came into production in the flourishing province of North Africa. Gaul and Germany became great wine-producing areas for the first time in history. The picture of agriculture in the early Empire seems to be a well-balanced one. There was room both for huge estates called latifundia, ranches, and specialized plantations and for the smaller market farms nearer cities. The disasters which destroyed Roman agriculture's ability to support the population of the Empire will be discussed in the chapter on the fall of Rome.

Roman farming developed considerable variety in its long history, within the limits set by the Mediterranean environment. Small farms were dominant during the earlier Republic; specialized plantations of grain, olives, fruit trees, or vines appeared, as did suburban farms raising fruit, vegetables, milk, eggs, poultry, and flowers for the markets of the nearby cities, and there was a tendency for farms to be amassed by the wealthy into latifundia.

Most of the Roman agricultural writers advise the careful siting of farms, with an eye to the variety of soils and exposures. Because of danger and the scarcity of water, most Roman farmers did not live on their farms; the small farmers lived in towns which were built on rocky elevations to save arable land, avoid the marshes, escape floods, and provide defense. Landlords generally lived in the city. But isolated farms and villas were known, particularly during the long periods of peace provided by the Roman Empire at its height.

The size of Roman farms in early times was small, as the population was very dense. About 4½ acres (7 Roman *jugera:* a *jugerum* is about .625 acre) were considered adequate as a plebeian allotment in 289 B.C. Colonial allotments in various parts of Italy which had been conquered by Rome were somewhat larger, and the Gracchan land reforms of the second century B.C. alloted about 19 acres per citizen-farmer. Roman land allotments were made not in a haphazard fashion but in a carefully surveyed rectangular grid marked out by roads, dividing the land into squares of about 125 acres (200 jugera). Begun in the third century B.C., this "centuriation" spread eventually over thousands of square miles in lands as disparate as the Po River Valley and North Africa. Such a rigid system, a forerunner of the American public land survey and the homestead law, could not easily be adapted to local conditions and betrays an exploitive attitude toward the natural environment. Today, Roman centuriation can be traced easily in aerial photographs of many areas in Italy.

The Roman agricultural calendar was not much different from that of the other Mediterranean peoples. Grains (wheat, barley, millet, and others) were planted in the autumn, grew slowly through the winter, matured and were harvested in the late

17. A rare example of relatively undisturbed Mediterranean vegetation, Mount Athos has been a monastic preserve for more than a thousand years, with sheep and goats excluded. Many species of flowers survive only here.

18. Spruce trees in one of the few forest remnants near Athens look down from Mount Parnes on rocky, eroded slopes like those described by Plato over 2,000 years ago.

19. The trees in the foreground have been protected as part of a religious shrine in the Apennine Mountains near Assisi, Italy, but the upper slopes have been subjected to deforestation and severe overgrazing.

20. The course of the River Peneios, with its braided channels at the upper end of the alluvial plain of Thessaly, shows some of the effects of erosion in the mountains from which its headwaters flow.

spring or early summer. Grapes, figs, and olives ripened from the late summer or into the autumn. Summer crops of fruits, vegetables, and fodder required irrigation, owing to the dryness of that season.

Hard as it may be to imagine Italian cooking without the tomato, it must be remarked that the variety of fruits and vegetables in Rome was small. The tomato, unknown in ancient Rome, was introduced much later from South America. Vegetables known to the Romans included beans, peas, lentils, turnips, radishes, carrots, cabbage, and lettuce. In their orchards grew apples, pears, almonds, and hazelnuts. This list was expanded by plants brought in from other parts of the Roman world and by trade with the East. Apricots and peaches came from the borders of Persia, and Lucullus is credited with bringing the cherry from Cerasus on the Black Sea. Other introductions of the Roman period include walnuts, mulberries, pistachios, and some kinds of melons. Introduced plants have an important effect in the ecosystem in replacing native species, and many cultivated plants can escape to the wild. Weeds were introduced also, though their arrivals were unintended and unchronicled.

New breeds of animals were also sought and introduced by the Romans. On their farms, they kept cattle, especially oxen used for plowing. All of the animals mentioned in the section on grazing could sometimes be raised on the farms, especially swine. Poultry of various kinds were known and raised by this time. The Romans also practiced the art of beekeeping, which provided their tables with sweetening and must also have had a generally good effect within the environment by providing pollination for many plants.

Cicero once remarked that "the farmer keeps an open account with the earth,"[6] which returns either a low or a high rate of interest. The Romans were aware of declining soil fertility and of some measures which could be taken to restore the soil. A literary tradition exemplified by Lucretius, Virgil, Ovid, and Seneca repeated the theme that the earth, once fertile, is becoming infertile through a natural process of aging, an irresistible advance toward senility. In the face of that idea, so strongly entrenched in poetry and prose, some astute Roman agricultural writers maintained that the observable decline in soil fertility was due not to the senescence of nature, but to abuse of nature by mankind. Misuse of the earth through bad agricultural practices, said authors like Columella and Pliny, causes crop failures and is attended by erosion. Still, many of the soils of Italy are durable and will recover with proper care.

Erosion, a constant problem on Mediterranean hillsides once the natural cover had been removed, was partly countered in some areas by terracing the slopes with stone walls, and by carrying soil back up once it had washed down.

The Roman farmer knew several methods of retaining and restoring soil fertility, including letting a field lie fallow in alternate years, and plowing repeatedly. He knew a form of crop rotation, using legumes to restore the soil in the year after it had borne grain, and even a three-field system in which grain, legumes, and fallow followed one another in successive years. But the actual use of crop rotation was restricted by the highly specialized utilization of the land, which virtually dictated that a given crop be grown year after year on the particular land best suited to it. Of course, intercultivation was widely practiced; wheat could be sown thinly among the olive trees. One might have seen the Mediterranean triad in one field: grapevines supported by olive trees with grain growing between the rows.

Romans fertilized the earth with animal and human manures, including the blood of slaughtered or sacrificed animals. Vegetable refuse was put on the fields, or green crops were plowed under to enrich the soil. The ground cover was burned—a rather wasteful method—or ashes were spread over the fields. Mineral fertilizers such as marl and lime were known and used.

The Romans increased their agricultural yield through the careful selection of seeds, an art in which they excelled. With so many methods of sustaining production, it seems surprising that Roman agriculture eventually failed to support the Roman population. But the possession of techniques does not mean that they will, or can, always be used.

Roman works for irrigation and drainage were very extensive, and they are deservedly famous. In building dams, canals, tunnels, and aqueducts they were preceded by the Etruscans and Carthaginians, but the Romans developed water diversion and control to a far greater extent. Water from the melting snows of the mountains, from numerous springs and streams, was conducted to the fields. Among the many aqueducts constructed by the Romans, those with the clearest water were used for drinking, those carrying less pure or turbid waters for irrigation. Underground cisterns and reservoirs were constructed for water storage. To raise water to the fields, the Romans possessed pumps, waterwheels, and the familiar shadoof, or bucket on a counterbalanced arm. They drained lakes and marshes through tunnels and canals to open up more arable or grazing land, while at the same time destroying these bodies of fresh water and wetlands rich in fish and wildlife. Irrigation brings with it the danger of saline and alkaline accumulations in ill-drained soil, particularly in warm climates and soils like those of the Mediterranean. The Romans do not seem to have been much aware of this problem and its cause, although they did notice cases of extreme sterility due to mineral deposits in the soil, as in North Africa.

The Roman government saw the regulation and encouragement of agriculture as one of its major tasks. Certainly almost all of the major works of water control and distribution were constructed by the state, and stringent laws protected them and governed their use. Theft of water and pollution of the aqueducts were severely punishable. Further, the Roman government offered tax relief for putting abandoned or unused land under cultivation, and its officials in the provinces encouraged maximum agricultural development, especially since this made their own exactions more profitable. Government policy did not prevent disasters, of course, and may even have produced some. A fairly steady supply of grain for the crowds in the city of Rome was obtained in various parts of the Empire at widely fluctuating prices. Serious droughts occurred at irregular intervals, bringing attendant famine, and food shortage was always a possible danger under the Roman Empire.

The Romans took measures to prevent erosion on agricultural land, but erosion continued anyway. Recent studies have shown that in central Italy, erosion rates in ancient times on land under human occupation were about ten times the average rate on undisturbed land. Ancient authors noticed that streams draining cultivated areas carried much more silt than streams which drained untilled watersheds. The loss of good soil through erosion must have had a serious effect in causing a decline in Roman agricultural productivity.

Roman cities clearly illustrate Roman organization of the environment. Wherever they had the opportunity to build towns and cities on relatively open sites, the Romans revealed their conception of order by imposing upon the earth a rigid, rectilinear grid of streets. Cities founded by the Romans all the way from North Africa to the outlying provinces of the north reflect a similar plan, basically square or almost square in shape, surrounded by fortified walls which were usually pierced by four gates, one on each of the four sides. Two major roads, the *decumanus* and *cardo,* connected the gates on opposite sides and met each other at right angles in the enclosed space. The other streets ran parallel to the *decumanus* or *cardo* and divided the city into rectangular blocks. Public buildings were located in places dictated by their importance, usually near or around a square where the two major streets crossed. This conception of a city developed from Greek ideas of city planning with a strong influence from the pattern of Roman military camps and fortresses. In fact, some of these cities actually began as permanent fortified military camps.

Vitruvius, the great Roman architect, theorized that cities should be built on dry, healthy sites and constructed so as to avoid and control the winds. His ideal city would have been octagonal rather than rectangular but would have retained most of the other elements found in existing Roman cities. To an even greater extent, Vitruvius's city, which was never built, would have imposed a rigid rational conception upon the chosen site.

A look at the city of Rome, home of the orderly and organizing Romans, reveals an almost complete lack of the neat planning visible in their colonies, and a violation of many of the principles set forth by Vitruvius. A network of crooked streets wandered among and over the famous seven hills and extended out past the irregular courses of the successive walls. Rome had not been planned originally, and efforts by consuls and emperors to bring order out of urban chaos met only with partial success.

Rome was built beside the Tiber River among hills, forests, and marshes in a "pestilential region," as Cicero and other Romans admitted.[7] There were several altars to the goddess Fever in Rome, with good reason. Various projects for drainage and sanitation were undertaken, but the relative unhealthiness of Rome remained a problem.

The physical growth of Rome through history as a group of agricultural villages, then a city-state, and later the capital of a huge empire was remarkable. Reliable figures on total population are not available for ancient Rome at any stage, but archaeological evidence indicates a large extension of the area covered by the city, and under Trajan, Hadrian, and Antoninus Pius in the second century A.D., the population may have reached 1,200,000. Roman writers comment on the suburban sprawl of the city, with villas of the rich occupying the nearby hills; as Horace said, "luxurious buildings leave few acres for the plow."[8]

Rome was very crowded, and the narrow streets impeded traffic, so endangering the unfortunate pedestrians that a municipal law under Julius Caesar prohibited all wheeled vehicles in Rome between sunrise and two hours before sunset, except for essential public service traffic; no doubt there was quite a rush hour between what corresponds roughly to 4:00 and 6:00 P.M., especially since the streets were not lighted at night. Modern efforts to close parts of central Rome to traffic have an ancient precedent. This law fell into disuse during the third century A.D. Juvenal and other Roman writers comment bitterly on the level of noise pollution in Rome generated by the traffic, industrial activities, building and wrecking, shouts of drivers and salesmen, and even the schoolchildren reciting their lessons in unison at the tops of their voices.

The houses of the wealthy in Rome, like their counterparts at Pompeii, were pleasant places, facing inward and insulated from the streets, with pools and gardens in their courtyards and perhaps even indoor plumbing using water from an aqueduct. But the great majority of Romans lived in insulae, large apartment buildings which were often too tall for their foundations and their supports. Augustus Caesar set a maximum height limit of about seventy feet for private buildings, but collapse was always a danger even so. There was no running water inside; the tenants had to use a public fountain down the street, and also one of the public latrines, or a commode. These last at times were emptied out upper-story windows. No glass or screens kept insects outside the apartments, and dust, dirt, and rubbish tended to accumulate inside because of the lack of plumbing. Since no fireplaces or chimneys were provided, charcoal braziers had to be used for heating and cooking, and oil-burning lamps for light; the smoke was supposed to blow out the window. Fires were common, and not easily extinguished or contained once they took hold in the crowded city. The famous fire of Nero was only one in a long series of conflagrations which burned whole sections of Rome. Under Augustus, a brigade of some seven thousand freedmen was formed to deal with these outbreaks, but it was only partly effective.

Air pollution was familiar to the Romans, who commented that the sun was so obscured by the smoke and dust that people coming back from the countryside would lose their tan in Rome after a few days. The countless cooking and heating fires and smoky lamps, the potters' kilns and bakers' ovens, the furnaces and hypocausts of the large and numerous baths all added their smoke to the dust raised by countless human activities. A trip out to the country, in the right direction, offered welcome relief.

The longing for the countryside, simple pleasures uncorrupted by the city, and beautiful natural things like trees, flowers, and fountains is expressed strongly in Roman writers such as Horace. It led the Romans to create an impressive series of private and public gardens to improve the quality of life inside the city. Large city houses and villas had gardens in their interior peristyles, often complete with pools and fountains. The poor had to be content with window boxes or small roof gardens. Public gardens surrounded by colonnades eventually stretched across the Campus Martius and other sections of the city. Roman gardens were highly formal, consisting of geometric lawns and flower beds, polygonal pools of water, symmetrical fountains, and waterfalls, with trees and shrubs clipped into fantastic shapes by the topiary's art. They constituted a major and successful attempt to make nature conform to the preconceived patterns of the human mind.

The original water supply of Rome was taken from springs within the city, from wells, and even from Father Tiber himself, Rome's river. But the latter two sources became notoriously polluted and unpalatable, and supplies were inadequate by 312

B.C., when Appius Claudius completed the first aqueduct.

Roman aqueducts remain one of the surviving wonders of the ancient world, even in a nonfunctioning condition, and they were certainly a source of pride to a Roman like Frontinus, who boasted, "With such an array of indispensable structures carrying so many waters, compare, if you will, the idle pyramids or the useless, though famous, works of the Greeks."[9] By gathering water from springs, lakes, and streams for many miles over the region of Latium, the Roman aqueduct engineers made a major impact upon the environment. It has been calculated that when all the aqueducts of Rome were functioning, they delivered to the city a flow of water which was at least one-third greater than the average flow of the Tiber River where it enters the city. The purity of water in the aqueducts was maintained by covered channels and the use of reservoirs and settling basins. There were strict laws against pollution and water theft of various kinds. Vitruvius's warning against lead pipes went unheeded, but deliberate pollution aroused public anger, as when Nero bathed ostentatiously at the intake of the Aqua Marcia. Some of the aqueducts, because they carried water unfit for drinking, were used for other purposes. The Romans knew the taste and temperature of the water from the various aqueducts, and had definite preferences. Water from the aqueducts was distributed to public fountains, to some industries such as textile fullers and dyers, to other public uses including the *naumachiae,* or sea fights, public entertainments in which boats manned by armed crews battled each other, to some private houses, and to the baths. The baths, and there were scores of them, provided daily diversion for the Romans, using prodigious quantities of water. The reservoir for the baths of Caracalla alone could contain 2,688,000 cubic feet. While springs provided naturally hot water to residents of the Bay of Naples and other places in the empire, at Rome furnaces raised water to the desired temperatures and heated the hollow floors and walls of many large rooms in the baths. Smoke from these furnaces and heating chambers, or hypocausts, was carried up to the roof, where it poured out through flues.

The excess water from the aqueducts was used to remove sewage from the city through the large, extensive system of sewers, which also served for storm drainage. During the Empire, there were spacious public latrines in Rome, richly decorated with marble and mosaics and seemingly designed for conversation as well as their primary purpose. (The emperor Vespasian met with some criticism when he taxed public conveniences—perhaps history's first pay toilets.) The sewers connected with the latrines, and with some of the larger private houses, but not usually with the insulae. Human excreta were often sold as fertilizer to farmers and gardeners.

The Cloaca Maxima was the largest sewer in Rome, but was only one of many. About fourteen feet high and eleven feet wide in some places, it could be maintained by workmen from within. Under ordinary conditions, much of Rome's waste matter was flushed out through the sewers and into the Tiber, which carried it down to Ostia and the Mediterranean. There was, of course, no sewage treatment of any kind. During floods, the Tiber often backed up through the sewers and inundated lower sections of the city. It is reported that the drain in the floor of the Pantheon looked like a fountain at such times, and the danger to health can be imagined. Also, the sewers, with their large, unguarded openings, harbored rats and other vectors of disease.

Under the Republic, the dead were usually buried in the ground. This was not permitted inside the sacred limits of the city itself, but these limits did not include all the inhabited area at all times. Bodies of paupers and others were simply thrown into open charnel pits. At the time of Augustus Caesar, efforts were made to end these practices. Cremation became more general, and one of Augustus's friends, Maecenas, covered over the worst area of the charnel pits and planted a garden on top.

Given these conditions of public health, it should not be surprising that Rome suffered epidemics of communicable disease from time to time. It is true that Rome had the best public sanitation of any large ancient city, but even in the days of greatest prosperity and material comfort under Marcus Aurelius, a serious plague swept the city and the entire Roman world. Hospitals in the modern sense of the word did not exist in Rome, and the ill were simply cared for at home, with a physician in attendance if the family could afford it. Temples of Aesculapius, god of healing, offered some psychosomatic help, and after the second century A.D., there were a few public infirmaries for the indigent sick.

Rome grew in population during some periods of her history, but it seems that the major source of increase must have come from migration to the city from the countryside and the provinces. Evidence is strong that such migration was the rule, and the lack of adequate sanitation and public health facilities, combined with prevalent infanticide and other problems, makes it appear that Rome, in spite of her imposing public buildings and material wealth, did not provide an environment within which mankind could flourish without a constant source of numerical replacement from outside.

Beyond the walls of Rome in every direction stretched the Roman roads. They are one of the most significant ways in which the Romans left their mark upon the landscapes of Europe, Asia, and Africa. The construction of the roads themselves is remarkable. Built with foundations secure enough to have supported a wall, they were carefully paved with stone, and followed purposeful lines, often straight across country, crossing marshes on causeways and rivers on magnificent bridges, many of which still exist today. Roman roads did not lie lightly on the countryside, nor were they carefully engineered in the modern sense of following contours, limiting grades, and avoiding unnecessary erosion, although they were located so as to be well drained. The Roman engineer was able to respond brilliantly to the challenges of narrow, rocky gorges and difficult mountain passes, but in gentler country he simply made his roads as straight as conveniently possible. In this, as in so many other ways, the Roman attitude toward the natural world is shown to be one of conquest and control of nature and confidence in the power of man. But the impact of the roads went far beyond their admittedly impressive physical construction. In a sense, they represent the octopoid tentacles of the city, reaching out to find subsistence in its surroundings. They hastened and secured the spread of Roman domination and made land transportation more convenient, rapid, and economical, and somewhat more competitive with sea transport. They encouraged the development of agriculture, mining, and industry farther from the metropolitan center by providing access to more distant areas. Because of them, more forests could be felled, and plants and animals were transported and extended their ranges, through deliberate or accidental introduction. The roads increased human mobility and reduced the inaccessibility of more distant areas, both factors which amplified the impact on the natural environment of the Romans and those who followed them and continued to use the same roads.

Roman military forces played a major role in the subjugation of the environment, as well as the people, in the vast provinces which they added to the empire. The army was active in peace as well as war after the first century B.C. Every Roman legion had its contingent of engineers and trained construction workers, and many major works were constructed with military labor. Camps were almost always fortified with ditches, embankments, palisades, and even walls. Forts often became the nuclei for towns and cities, particularly in the north and west. A huge organization supplied the army and navy, using large warehouses in the major cities, ports, and fortresses, and pack trains along the roads and trails. Roman law strictly forbade the sale of any provisions which were destined for the Army. Infantrymen were issued grain, wine, vinegar, and even salt in specified quantities. Meat and cheese were often provided; inscriptions indicate that some soldiers were designated to guard cattle attached to the army. Pasture or fodder was required every day for the cattle, cavalry horses, and pack mules that accompanied a Roman legion. Military law described pack animals as more important than weapons, because they are necessary all the time, not just in battle. Their demands depleted the grazing lands in areas where military units were stationed. In addition, official hunters were assigned by the legionary commanders to scour the surrounding woods for boars and deer to augment the army's diet. The army also was detailed to capture wild animals for the arena.

Since the army was carefully supplied, military law sought to prevent it from living off the land by seizing supplies within Roman territory. During the conquest of enemy lands, foragers could be sent out to find provisions wherever they could. Trajan's column bears a relief showing Roman soldiers cutting grain in Dacia. In times of dire military necessity, commanders would requisition supplies from neighboring villages, as Caesar did in Gaul. Control and discipline sometimes broke down, as they did in the period of sporadic military uprisings in the third century A.D. when the contending armies of rival candidates for the imperial throne marched back and forth across the Roman world, ravaging farmlands and taking whatever they wanted. These unsettled times, longlasting and recurrent, took their toll of the Roman agrarian economy, particularly through the drafting and slaughter of the farm population.

Destruction of human life and devastation of the natural environment went hand in hand in Roman warfare. The historian Tacitus has a British chieftain say of the Romans, "They make a desert and call it peace."[10] Roman generals attempted to deprive their opponents of every means of subsistence and, therefore, resistance. While they were often generous to those they conquered, they were capable of leveling enemy cities to the ground, as they did at Corinth. After destroying Carthage, the Romans ordered the fields sown with salt so that nothing would ever grow there again. A hundred years later, the salt had already leached from the soil when Julius Caesar commanded that a new city be built on the site. Of course, military operations do not always deplete the soil. After the battle of Massalia, in which Marius defeated the Teutons, Plutarch reports:

> It is said that the people of Massalia fenced their vineyards round with the bones of the fallen, and that the soil, after the bodies had wasted away in it and the rains had fallen all winter upon it, grew so rich and become so full to its depths of the putrefied matter than sank into it, that it produced an exceeding great harvest in after years, and confirmed the saying of Archilocus that "fields are fattened" by such a process."[11]

11

Ecology and the Fall of Rome

The Romans' failure to adapt their society and economy to the natural environment in harmonious ways is one of the causes of the decline and fall of the Roman Empire, if not in fact the basic and underlying one. The Romans placed too great a demand upon the available natural resources, depleted them within their sphere, and failed to maintain that balance with nature which is necessary to the survival of any community. Of course, the fall of Rome was a large and complex phenomenon that cannot be attributed to a single cause. Ecological failures interacted with social, political, and economic forces to assure that the vast entity called the Roman Empire would disappear or be changed beyond recognition. Unfortunately, the ecological factors have not received much attention from historical and classical scholars. They have been dismissed as unproven without thorough consideration and investigation. It is to be hoped that in the next few years, historians of Rome will turn to the discoveries being made by scientists which bear upon these problems, and come to a new appreciation of the importance of environmental changes in the fall of Rome. "When Rome falls, the world falls, too," ran an old saying quoted by Gibbon and Byron, and it may be time to consider the converse, for when Rome made nature a slave and tried to work that slave beyond endurance, the natural world "fell," or at least lost the ability to support the mistress of the world, and Rome fell, too.

Strangely, the characteristics and achievements of Rome that modern Americans and Europeans tend to admire most seem to be the very things that magnified and hastened the depletion of the natural environment and helped to make the fall certain. Massive structures of many kinds simply underline the ability of the Romans to alter nature, and the spread of Roman commerce throughout the Mediterranean and parts of the three adjoining continents made their impact on nature that much more extensive. The utilitarianism of the Romans led them to treat the natural environment as an apparently inexhaustible resource rather than as a living system, and their love of rule encouraged them to regard nature as conquered territory.

Modern scholars who have discussed possible environmental reasons for the fall of Rome have sometimes failed to distinguish between two interrelated but separate sets of causes: those that are produced by the activities of mankind, and those that are not. Forces which impinge on mankind from the environment without apparent human cause, such as climate and epidemics, are not yet fully understood. But it may still be true that they have influenced the rise and fall of nations, including Rome.

Changes of climate in historical times are undeniable. The round of seasons is never exactly the same in two different years, and periods of unusual heat and cold or drought and flood have always been noticed by mankind. The shorter fluctuations of rainfall and temperature are more easily noticed, but there are long-term changes of averages over periods of centuries and millennia. Before the rise of agriculture, the latest in a series of ice ages came to an end. Changes since then have not been so extreme, but they have happened. As an example of evidence of climate change in historical times, the sea level is now higher than it was in classical Greek and Roman times, probably due to the partial melting of the polar continental ice sheets by higher average temperatures. Scholars would like to have a reliable chart of the changes in climate that have occurred in different periods of history, but although science may provide such a chart in the near future, it is not yet possible to talk about climate changes with a close degree of chronological precision. The general pattern of change during the Roman Empire, from the somewhat equivocal evidence presently available, seems to have been toward moister conditions in the earlier period, but warmer and dryer in the later period. The ancient Greeks appear to have flourished during one of the colder and moister episodes, and the fall of Rome occurred during a warm, dry phase. The present state of knowledge does not permit too much speculation beyond these generalizations, and it should be emphasized that these long-term changes in climate during historical times in middle latitudes have not been catastrophic or of severe magnitude. During a period when the average temperature is a degree or two warmer, a very cold winter can occur, and droughts can occasionally happen in centuries which are, on the average, wetter. Short-term changes over a period of years, like floods and famine-producing droughts, are always more noticeable than long-term ones over periods of centuries. Within the next few years, it may be possible for historical climatologists to say much more about climate change in the period of the Roman Empire. It is impossible to rule it out as a contributing cause in the decline and fall of Rome. One possibility, for example, is that changes affecting the growth of grass in Central Asia may have started nomadic barbarians outward in waves of movement which eventually broke on the frontiers of the Roman Empire. Whatever climate changes may have occurred, however, it seems quite clear that agricultural deterioration was chiefly the result of human mismanagement and neglect.

Severe outbreaks of communicable disease have periodically attacked mankind. Without modern methods of inoculation, ancient populations developed immunities at great cost in human life, but new mutations among disease organisms produced new epidemics. Without doubt, plagues have affected history. The Spartans won the Peloponnesian War partly because Athens had been weakened by plague. In Rome, a plague broke out during the reign of Marcus Aurelius, killing as much as a third of the population in some areas. Human numbers tend to build up after such attacks, but the loss may not be repaired for many decades even if other disasters such as wars and new epidemics do not intervene. In Rome, there was no respite in the generations following A.D. 180. Thus disease appears to have played a role in reducing the population of the Roman Empire.

Many of the ecological factors that contributed to the fall of Rome were the result of human activities. In these cases, it could be said that the Romans might have helped save their society, or at least delayed or softened its fall, if they had seen what they were doing and taken action to modify it.

It was certainly evident to the Romans that the forests were being removed, because vast tracts of former forest land had been denuded within their memory. But they tended to see this process as inevitable and even advisable, since it made land available for grazing and crop growing. The first few crops on recently cleared soil were usually quite rich, owing to the humus and minerals left by the forest, so the loss of the available timber seemed to be amply repaid in the new use of the land. But erosion began immediately, particularly on sloping land, and the earth yielded less fruit every year until, in some cases, all the soil had been washed away.

The Roman government treated forests as a nonrenewable resource, handing them over to publicans of the equestrian class for exploitation. Syndicates were formed to strip the forests from the hills and sell the timber for profit as rapidly as possible.

The effects of deforestation were widespread. The changes made in the larger patterns of temperature and rainfall are conjectural, but the effects on wind, temperature, and humidity in the localities actually deforested were certainly great. Extremes of heat and cold were greater without the moderating presence of trees. Winds blew across the ground without much resistance. Without the interlacing branches and root systems of the forests to intercept rainfall, the runoff increased. Erosion by wind and water was accelerated over millions of acres. The removal of the bulk of the Mediterranean forests was, and remains, the most damaging and visible evidence of human activity.

Exhaustion of the more available sources of metal was a definite problem for the Romans, and they were forced to seek sources of supply in mines located at greater and greater distances. The problem was especially acute in regard to the precious metals, since gold and silver were the mainstays of Roman coinage. Importers of silk, spices, ivory, and other luxuries from India, China, Southeast Asia, Arabia, and East Africa, the Romans had little to offer in return except their gold and silver. The balance of trade was definitely not in their favor. Pliny the Elder thought that the Roman Empire lost about 100 million sesterces every year to her eastern neighbors, and caches of Roman coins found in various corners of Asia and Africa give some support to his statement.[1] During the later centuries of the Empire, Rome suffered ruinous inflation and her coins were repeatedly debased. Alloys were made with less of the precious metals, or coins of base metal were thinly jacketed in silver. The government often paid salaries in bronze or in kind, but tended to demand certain taxes in gold and silver. These symptoms all seem to point to the progressive exhaustion of the mines known to the Romans, which forced them to resort to practices more damaging to the environment and more dangerous to their own lives and health. Pits were dug deeper, poorer lead-silver ores were smelted, and more forests were cut to be used in the mines and furnaces.

The extent of lead, mercury, and arsenic poisoning among the

Romans, closely associated with mining and purifying the precious metals and other industrial processes such as the working of leather, textiles, and pottery, is as yet unknown. Investigations are now being undertaken to determine the concentrations of lead in bones which have been recovered from Roman burials. In the form in which it is generally advanced today, the argument in favor of lead poisoning as a major factor in the fall of Rome holds that the upper classes, and therefore the intelligentsia, of Rome were particularly exposed to lead in the form of utensils, dishes, cooking pots, and jams and sweeteners containing a high concentration of lead compounds. The effects of lead poisoning, including interference with reproduction, physical weakness, and dulling of the intellectual faculties, were cumulative, long-lasting, and not easily seen to be connected with the cause. But it is not necessary to insist that lead poisoning acted only, or even primarily, on the higher social classes in order to see it as a factor in the fall of Rome. One of Rome's lasting virtues was her ability to fill the ranks of an enfeebled and shrinking ruling class with vigorous, intelligent people from the lower classes and from other nationalities within the Roman sphere. Even slaves, or ex-slaves, rose to positions of leadership. If Roman technology provided the population in general with debilitating industrial poisons through the aqueducts, in their diets, in the atmosphere, and in the commonly handled objects of daily life, this in itself could be shown to be a cause contributing to the fall of Rome.

Overgrazing was a widespread force of environmental degradation in Rome. Large areas were converted to this use from forests and even from croplands. Overgrazing hastened deforestation and made it permanent through clearing, the use of fire, and the destruction of tree seedlings by grazing animals. With overgrazing came serious erosion. There are sections of Italy where the hillsides, originally forested, are essentially barren, although the alluvial land in the valley bottoms is still fertile.

Here the problem since Roman times has been not simply deforestation, but the annual destruction of the grasses and herbs which would otherwise have begun a natural succession on the denuded land. For the Romans, the result was to decrease the available productive land that formed the base of their essentially agrarian civilization, and thus to reduce the number of people who could be supported within the Empire.

Soil exhaustion of arable land was an important cause contributing to the failure of Roman agriculture and the fall of Rome. It is a cause which has received too little attention in recent years. Roman agricultural writers, with the outstanding exception of Columella, tended to see the exhaustion of the soil as an inevitable result of the aging of the earth, and the noted modern proponent of soil exhaustion as a cause of Rome's fall, Ellsworth Huntington, saw it as an accompanying result of climate change and therefore beyond the control of mankind. Thus soil exhaustion has come to be associated in many minds with an environmental determinism that is hard to defend owing to lack of evidence. But mankind can enrich the soil through care, and most soil exhaustion is the result of human acts of omission and commission. As Rostovtzeff remarked,

> If, therefore, there was exhaustion of the soil in Italy and in the provinces in the centuries after the great crisis of the third century, this must be ascribed to man, not to nature. Men failed to support nature, though they knew as well as we do, or as the Japanese and the Chinese how it should be done. It is very probable that, in the late Roman Empire, exhaustion of the soil in some parts was a real calamity.[2]

Although the great economic historian is arguing here against regarding soil exhaustion as a primary cause of the decay of Rome, his argument clearly supports the contention that human "failure to support nature," for whatever political, economic, or military reasons, did affect the ability of the earth to support the

Romans, and therefore forms an important link in the chain of causation. Human neglect of nature could be felt even in Egypt, where the Nile renewed the soil almost every year. Egypt might easily recover, but in other areas the problems of irrigation and drainage, salinization, and fertilization would have been acute. Land was progressively abandoned from the third century A.D. onward, when Roman documents make increasing references to *agri deserti,* deserted fields.

In the natural state of the Mediterranean landscape, erosion would for the most part have occurred slowly. Soil was built up through millennia. But with deforestation, overgrazing, and plowing to remove the vegetative cover of the earth, the soil was exposed to the action of wind and water, and was carried away far more rapidly. This produced a continuing crisis of declining soil fertility in the Roman Empire. While some of this soil was deposited on the valley floors and other places where it could be used, more of it was carried down the rivers into the sea, where it added measurably to the land area, but created swamps which were difficult to reclaim. Malaria, carried by the mosquitoes which bred in these swamps, was a long-lasting disease which usually was not fatal but did help to weaken the Roman Empire by sapping the energy of its citizens.

For several reasons, agriculture failed even where, with proper care, the soil would have been perfectly capable of producing adequate crops. During the third century A.D., the almost constant military campaigns of various claimants for the imperial throne against each other devastated the countryside. Farmers, always a major source of military manpower, were forced into service or slaughtered. Armies lived off the land. Small farmers were ruined and tended to become dependents of great landowners, as large estates survived more successfully and increased again in the period which ensued. The system of taxation encouraged the depletion of the land. From the time of Septimius Severus in the early third century, a fixed annual tax, the *annona militaris,* was collected in kind, and did not vary with the yield of the harvest. This and similar taxes, used in large part to support an ever-growing number of nonproducing people in the army and bureaucracy, were collected chiefly from the agricultural sector of the economy.

A chronic food shortage as the result of agricultural decline seems to have resulted in a declining population for the Roman Empire during most periods after the second century A.D. Of course, declining population also meant fewer farm workers, so the reductions in agricultural production and population tended to reinforce each other. Diocletian's edicts on prices and occupations, which failed in their purpose, indicate what was happening at the end of the third century; food was becoming scarcer, prices were rising, and there was a shortage of labor. There were several causes for the decline in population besides the agricultural crisis. The wars of the period of military anarchy in the third century and the dynastic wars of the fourth, the violent incursions of barbarians across a less well-guarded frontier, and periodic outbreaks of plague all had their effect. In an urbanized world, conditions of city life did not favor an increasing population. Infanticide was widely practiced before the Empire became Christian, and there were few incentives to encourage raising children.

In the West, sections of the countryside were depopulated, and city walls were rebuilt to enclose smaller spaces. With a more dependable food supply in Egypt, geography less open to barbarian invasions, and something of a population reservoir in Asia Minor, the eastern half of the Empire suffered less from depopulation.

From what has been said, it should be evident that the most important ecological factors in the fall of Rome were those due to changes made in the natural environment by mankind, although factors beyond human control may also have been influential. Rome's treatment of nature helped bring about

Rome's decline and fall. But if this is true, then it might well be asked what it was in the relationship of the Romans to nature that led them to make serious mistakes, and prevented them from seeing and correcting these mistakes.

The characteristic Roman attitude toward nature was partly to blame. While the early Romans had worshiped nature and had been inhibited from making major changes in the environment by religious taboos, the Romans of the middle and late Republic and the Empire were increasingly utilitarian and willing to exploit their natural resources. From Greek sources, the Romans imbibed a rationalism which seemed to justify their own tendency to regard the natural world as a series of things instrumental to human desires. While Latin literature shows that the Romans never lost an aesthetic appreciation of natural beauty, this feeling was most characteristically directed toward landscapes controlled by mankind, and never interfered in such practical matters as mining, grazing, and logging. An environmentalist movement did not exist in Rome, unless it can be seen in the bitter reflections of Juvenal, Martial, and others on urban conditions and the habit wealthy Romans developed of retiring to country villas to breathe the pure air, escape the noise of the city, and feel close to the land. Roman urban planning did reflect the desire to create a pleasant and healthy environment, of course, but it did so by a kind of rigid structuring and triumph over nature which was never wholly successful. Stoic philosophy, long popular and even dominant at Rome, gave formal recognition to nature and natural law but applied its teachings to the inner life of the individual, to law and social relationships, not to human treatment of the environment. Many Romans believed that the environment influences mankind, but this belief took the form of a theoretical environmental determinism usually connected with astrology. Cicero, like many others, recognized that "by means of our hands we essay to create as it were a second world within the world of nature," that is, that mankind has a major impact upon the natural environment. But this was almost always expressed along with great pride in human ability, without any sense that damage might be done:

> We enjoy the fruits of the plains and of the mountains, the rivers and the lakes are ours, we sow corn, we plant trees, we fertilize the soil by irrigation, we confine the rivers and straighten or divert their courses.[3]

Columella was alone in warning that nature's failures to produce abundantly were due to human mismanagement, and he does not seem to have made much practical impression.

To the Romans, nature was comfortably familiar. She provided the materials and background for their achievements and offered a refreshing alternative to urban life. But nature was not seen as possessing a value in herself, nor was the relationship with nature ever seen as crucial to the survival of society and humanity itself.

The Roman failure in relationship to nature was also a failure in understanding. One need only turn to the *Natural History* of Pliny the Elder to see the misconceptions and myths about the earth, animals, and plants that misled the Romans, and Pliny probably represents the more enlightened people of his society. The practical minds of the Romans showed disregard for theoretical science, and such scientific research as was being carried on at Greek centers like Alexandria came to a slow and painful halt under the Romans. Pragmatic politicians and technicians are usually not friendly to the scientific researcher; his investigations seem to be of no earthly use while he is pursuing them, and when he discovers a useful application, it seems either self-evident, or revolutionary and threatening to society and the economy. Only rarely have governments or benefactors seen the value in maintaining free scientific inquiry, trusting that increasing knowledge will prove to have some value and even some practical application eventually. The

Hellenistic kings of Egypt did this to some extent, but the Romans did not do so at all. Emperors did endow chairs in rhetoric and philosophy. But even in such obviously useful fields as medicine and agriculture, virtually no scientific research was undertaken and no important theoretical discoveries were made. Since Roman science was stultified, there was no increase of knowledge about nature, and no development of ecological understanding.

Roman ability to affect the natural environment was the result of technology, and Roman technology was one of the wonders of the ancient world. Roman engineers were skilled in the use of large wooden machines in peace and war. They developed the use of concrete and the true arch and its derivatives in architecture. They harnessed the waterwheel, made advances in metallurgy, and invented harvesting and threshing machines for use in northwestern Europe, where agricultural conditions differed from those of the Mediterranean Basin. Granted the technical achievements of the Romans, it is hard to understand why they did not do more to develop sources of energy and use those they had more efficiently. The principle of steam power was known at Alexandria, where it was used to open doors and perform simple temple miracles. It was left to the barbarians to invent the stirrup, horseshoe, and heavy plow, and the successor states came up with the horse collar, which enabled horses to pull plows without choking, and the windmill, which utilized another source of energy. All of these inventions would have been useful to the Romans if they had thought of them. In point of fact, they could have had the printing press, gunpowder, and the compass as well. The Roman failure to invent and use labor-saving devices has been attributed to their dependence on slave labor. Why invent a machine or harness the wind when there are slaves to do the job? But it seems likely that if new methods had proved to be more economical than slavery, the Romans would have adopted them, and late in the Empire, there was a general shortage of laborers. Rome was characterized by a resistance to the new and a certain lack of imagination. It is reported that the emperor Tiberius, on hearing that a flexible glass had been invented, had the inventor's workshop destroyed and the invention suppressed because he feared a drop in the value of the precious metals. Vespasian rejected an inventor's column-moving machine because it might produce unemployment. The Roman government did not support technological advance. If the Romans had developed their technology further without much change in their attitudes toward and knowledge of nature, their impact upon the natural environment would probably have been swifter and more destructive. Some of the results, including an increase in agricultural production, might have been good. But if technology was to serve as a means of improving the relationship of mankind to the environment, rather than simply to magnify the ability to alter and destroy the environment, it would have had to be directed and controlled by ecological concern and knowledge. If the Romans had decided to do this, they might well have succeeded, for their system of government, laws, and social control in general was quite effective and persisted for centuries with relatively few breakdowns. But, as has been seen, the Roman attitude toward nature precluded such a decision, and their lack of ecological understanding in any case would have prevented them from knowing what steps ought to have been taken.

12

Christianity and the Natural Environment

Christianity appeared during the early Roman Empire, spread throughout the Mediterranean area and beyond it, and within four centuries had become the dominant religion and system of thought in the Near East, North Africa, and Europe. The influence of Christian attitudes toward the natural environment was widespread, particularly in the last few centuries of ancient history. Some of the central teachings of Christianity bore directly upon the relationship of mankind to nature, but probably just as important were the implications of Christian ideas concerning humanity and the environment which gradually became evident as thought and life developed in societies influenced by Christianity.

The most important Christian ideas about the natural environment were inherited directly from Judaism. Christians retained the Jewish Bible as the Old Testament and accepted its teachings concerning God's creation of the universe and providential rule over it, and human dominion over and stewardship of the natural world, with ultimate responsibility to the Creator. Therefore the description of Jewish attitudes given in chapter 5 applies also, in great measure, to Christianity.

One of the most striking characteristics of Jesus himself was his evident love of nature, exemplified in a life largely spent out of doors in the villages, countryside, and wilderness of Palestine. Most of the parables, similes, and metaphors in the recorded sayings of Jesus use images taken from the natural world, and references to trees, birds, seeds and growing grain, vines, sheep, and the work of farmers and herdsmen are common in the New Testament. "Consider the lilies of the field," said Jesus, "they neither toil nor spin; yet I tell you, even Solomon in all his glory was not arrayed like one of these."[1] Of course Jesus used the references to nature in his parables in order to illustrate truths about human spiritual life, not simply to describe nature. In order to discover the attitude toward the natural environment which is implicit in these parables, therefore, it is necessary to use a method which distinguishes the analogy or comparison which Jesus made, relating an image from nature to a moral or spiritual point, from the ideas concerning nature herself which are evident in the parable. For example, in the parables about the mustard seed, Jesus compared faith to a tiny seed which grows until it becomes a very large plant. The point of the parables is clearly the potential and increasing power of faith. But to make such a comparison possible and so singularly appropriate, it was necessary to observe the amazing process of reproduction and growth in plants, where a huge bush develops from a tiny seed that contained within itself the power and pattern for later growth. Many of the parables can be analyzed in this way to reveal much valuable information about the attitude of Jesus and his hearers toward the world of nature. It is possible to see in his many references to seeds cast into the earth and growing, "first the blade, and then the ear, and then the full grain in the ear,"[2] an appreciation of the order and regularity of the cycles of nature, a principle which also underlies his remarks about the signs by which farmers and sailors can tell the onset of seasons and changes in the weather.

Jesus often retired for prayer and spiritual renewal to the mountains, the sea, or the desert, which is called a wilderness or a "deserted place" in the Greek of the New Testament. He also chose such places to speak to his followers. He taught beside the sea, went up on a mountain to deliver a sermon, and invited his apostles to come apart with him "to a desert place, and rest a little."[3] Many of the most important events of his life took place in the wilderness; he was tempted and strengthened for his mission there, the transfiguration took place on a mountain top, and the painful decisions of his last week on earth were made on the Mount of Olives and in the Garden of Gethsemane. The chief characteristic of such places is their isolation; in contrast to city and village, they were quiet and free from the distracting presence of crowds of people, suitable places to go for a period of time to commune with God.

Jesus reminded his followers often of mankind's dependence upon the natural world. He advised them to pray for daily bread, and often told parables based on the occupations of those who gain their living by working with nature and its products, such as fisherfolk, farmers, shepherds, and vinedressers. He took the common elements of nature found in the environment of Palestine as sacramental substances, investing them with deep spiritual significance. He used the giving of a cup of water as a sign of one person's kindness to another. Water he regarded as both a literal cleansing agent—he advised a healed man to wash in it—and a sign of spiritual cleansing and rebirth in baptism. In the last supper which he shared with his disciples before his death, he gave particular meaning to bread and wine, the staple ingredients of the Mediterranean diet and elements also blessed by prayers in the Jewish tradition. Far from making a separation between the human spirit and the natural world, Jesus believed that people should take joy in nature's beauty, learning from the examples offered by nature and leading a simple life, accepting in gratitude the necessities of life from nature as gifts of God.

Christianity's central teaching about the natural world has always been that it is the creation of God. The New Testament speaks of "a living God who made the heaven and the earth and the sea and all that is in them."[4] As Creator, God continues his concern for his creation and provides for his creatures. It is "He who supplies seed to the sower and bread for food."[5] But his providence is not for mankind alone, for "look at the birds of the air; they neither sow nor reap . . . and yet your heavenly Father feeds them."[6] Not a single sparrow "is forgotten before God."[7]

The New Testament also repeats the Old Testament affirmation that the created world reveals the hand of the Creator to observant people. Paul, in a well-known passage in Romans, maintains that all people should be able to see the creative work of God manifested in the world. "Ever since the creation of the world his invisible nature, namely his eternal power and deity, has been clearly perceived in the things that have been made."[8] As the creation of God, Christianity teaches, the natural environment is in itself good. "Everything created by God is good, and nothing is to be rejected if it is received with thanksgiving."[9]

This good world, according to the New Testament writers, was given by God as a trust to human beings. Mankind is God's vicegerent, his representative or steward. "Now in putting everything in subjection to man, he left nothing outside his control."[10] All kinds of animals can be tamed or hunted, killed, and eaten by people. Human dominion is not necessarily seen as exploitive, however. It is the meek who shall inherit the earth, and people must make account of their stewardship to God.

But according to the Christian view, since created things are subject to mankind, they are subject to a weak and fallen species which is in need of reconciliation with God. Mankind has failed as God's steward. Being created in God's image, people have instead rebelled against God and attempted to make themselves the lords of creation. The natural world suffers in a state of subjection and decay, but when humanity is saved, the natural order is saved also.

> For the creation waits with eager longing for the revealing of the sons of God; for the creation was subjected to futility, not of its own will but by the will of him who subjected it in hope; because the creation itself will be set free from its bondage to decay and obtain the glorious liberty of the children of God. We know that the whole creation has been groaning in travail together until now; and not only the creation, but we ourselves, who have the first fruits of the Spirit, groan inwardly as we wait for adoption as sons, the redemption of our bodies.[11]

Jesus Christ appears in Christianity, therefore, not only as the Savior of mankind, but also as the Savior of the whole created world. Christ is identified with the Logos, or creative word of God, who "was in the beginning with God; all things were made through him, and without him was not anything made that was made."[12] Since it was through Christ that God created the world, Christ also bears the nature of God in providence, "upholding the universe by his word of power,"[13] for "in him all things hold together." Christ's power over the elements of nature was clear from his earthly life as described in the New Testament, in his ability to calm the wind and sea and heal the suffering bodies and minds of human beings. It is further through Christ that the broken harmony of the world is restored, because God wills "through him to reconcile to himself all things, whether on earth or in heaven."[14] Thus Christianity regards Christ as having the role of cosmic savior. Those who became Christians in the ancient world learned that their relationship to the natural world could be defined in these terms: "All things are yours, and you are Christ's and Christ is God's."[15] As in the Jewish view, people are in control of the world but are ultimately responsible to God for their actions in regard to it.

At the same time, Christians were urged not to "love the world or the things in the world."[16] They should worship God the creator, and not created things, which, good and important though they are, cannot claim the ultimate concern of people. While the natural environment reveals God's creative work and providence, it must not be allowed to stand in the place of God or as a barrier between people and God. These New Testament warnings are in actuality not directed against nature itself, which was created good and remains good, but against the rebellious state of mind of mankind, which is all too ready to turn away from God.

Christians were taught not that the natural world is without any intrinsic value but that it is in its present form temporary and insecure. "The things that are seen are transient, but the things that are unseen are eternal."[17] The imminent destruction of the world by a disaster which is at least in part a catastrophe of nature, in which the earth and seas and heavens "will all grow old like a garment," and "like a mantle" God will "roll them up,"[18] was described in graphic terms in the New Testament. But it should be pointed out that these events were to mark not the abolition of the natural environment but its transformation and the restoration of its original harmony. A more perfect cosmos will succeed the old; "a new heaven and a new earth," as the result of a new creation: "Behold," says the Lord, "I make all things new."[19]

13

The Ancient Roots of Our Ecological Crisis

The damaging changes being suffered today by the natural environment are far more rapid and widespread than anything known in ancient times. Today deforestation proceeds on a worldwide scale, the atmosphere becomes more turbid and opaque every year, the oceans are being polluted on a massive scale, species of animals and plants are being wiped out at a rate unmatched in history, and the earth is being plundered in many other ways. But although the peoples of ancient civilizations were unfamiliar with such recent discoveries as radioactivity, insecticides, and the internal combustion engine, they faced problems sometimes analogous to those the modern world faces, and we may look to the ancients in order to see the beginnings of many of our modern difficulties with an environment which is decaying because of human misuse.

A human community determines its relationship to the natural environment in many ways. Among the most important are its members' attitudes toward nature, the knowledge of nature and the understanding of its balance and structure which they attain, the technology they are able to use, and the social control the community can exert over its members to direct their actions which affect the environment. The ancient world shows us the roots of our present problems in each of these areas.

In a well-known and often reprinted article, "The Historical Roots of Our Ecologic Crisis,"[1] Lynn White traced modern Western attitudes toward the natural world back to the Middle Ages. But both medieval and modern attitudes have ancient roots. Greece and Rome, as well as Judaism and Christianity, helped to form our habitual ways of thinking about nature. And it is evident that the modern ecological crisis is to a great extent the result of attitudes which see nature as something to be freely conquered, used, and dominated without calculation of the resultant cost to mankind and the earth.

These attitudes stem from similar ideas which were held by the ancient peoples who have most influenced us. Animism, which saw the natural world as sharing human qualities and treated things and events in nature as sacred objects of respect or worship, was the dominant attitude in early antiquity and persisted almost everywhere in the Mediterranean world, but it gradually gave way to other ways of thinking. In Israel, transcendent monotheism replaced animism's "world full of gods." Instead of being divine in itself, nature was seen as a lower order of creation, given as a trust to mankind with accountability to God. But in the later history of that idea, people tended to take the command to have dominion over the earth as blanket permission to do what they wished to the environment, conveniently forgetting the part about accountability to God, or else interpreting most human activities as improvements in nature and therefore pleasing to God.

Perhaps even more important in the history of human attitudes toward nature was the departure from animism made by the Greek philosophers. Rejecting traditional mythological and religious explanations of the natural world, they insisted on the ability of the human mind to discover the truth about nature through the use of reason. Instead of a place filled with spiritual being, or beings, a theater of the gods, the environment was to them an object of thought and rational analysis. Worship of nature became mere ritual, supposedly replaced with philosophical understanding. Since, in the words of Protagoras, "man is the measure of all things,"[2] it followed that all things have usefulness to mankind as their reason for existence. This idea has persisted in Western thought in various forms until the present, for the belief that everything in nature must justify its existence by its purposeful relationship to mankind is firmly, though perhaps implicitly, held by most people.

What was for the Greeks a philosophical opinion became for the Romans a practical reality. Early Roman animism was overcome less by the ingestion of Greek ideas than by the Romans' own demonstrated ability to dominate and to turn most things to their own profit, but both Greek influence and Roman practicality helped the Romans to develop attitudes toward nature which are remarkably similar to those expressed and demonstrated today. The Romans treated the natural environment as if it were one of their conquered provinces. If they needed any justification of this beyond their own pragmatism and cupidity, they could find it in Greek philosophy, which reached them in a late, skeptical form that had removed the sacred from nature and made nature an object of manipulation in thought and, by extension, in action. Our Western attitudes can be traced most directly to the secular, businesslike Romans. Today the process of dominating the earth is seen not as a religious crusade following a biblical commandment but as a profitable venture seeking economic benefit. In this, we are closer to the Romans than to any other ancient people, and in this we demonstrate to a great extent our heritage from them.

Attitudes alone do not determine the way a human community will interact with the natural environment. People whose religion teaches them to treat the world as a sacred place may still manage to make their surroundings a scene of deforestation and erosion, because good intentions toward nature are not enough if they are not informed by accurate knowledge about nature and its workings.

The earlier civilizations of the Near East accumulated a vast amount of information about the world through trial and error, and that information was passed on through tradition. Some of what they thought they knew was correct and useful, and much was colorfully inaccurate, interwoven with myth and folk stories.

A few Greek thinkers were the first to approach the natural world in a consistently rational fashion, demanding that reasonable explanations be found for all natural phenomena. This enabled them to begin the process of gaining knowledge which eventually developed into what might be called the scientific method. Many of the Greek thinkers were also careful observers of nature and attempted to check their ideas against what could be observed, but all of them held rational thought to be superior to what could be seen in the world and assumed that the inner workings of the human mind are congruent with the outer workings of the universe. This assumption, along with the antipathy of Greek thinkers toward work done with the hands, limited the range of their discoveries and led them into some fallacious speculations. Nonetheless, the discoveries of the Greek philosophers and scientists are many and impressive.

Unfortunately, research and discovery in this field gradually diminished under the Romans, who were collectors of older bits of information about the world of nature rather than discoverers of new knowledge. A few Greek scientists continued to work under the Roman Empire, but the Romans themselves produced few creative thinkers in this field. With the advent of Christianity, the situation worsened. Living in a world which they believed to be temporary, early Christians seemed to regard study of the things of this world to be irrelevant, if not a positive barrier on the way to salvation. "The wisdom of this world is foolishness,"[3] said Paul. He spoke in a somewhat different connection, but the Christians of the later Roman Empire, with very few outstanding exceptions, tended to look at all scientific inquiry in the spirit of that statement.

The modern world, having revived the works of the ancient Greeks, has gone beyond them in developing a rigorous methodology for gaining knowledge about the natural environment. The extent and accuracy of the understanding of nature that is available today is truly impressive, but far from complete. Much remains to be discovered about the circulation of the earth's atmosphere, weather, and the effects of pollution of various kinds on climate, for example. The behavior of species of animals and the interaction of all forms of life in an ecosystem are only imperfectly understood. Governments and institutions have not always seen the relevance of such knowledge, and support for research has been a sometime thing. At the same time, human activities are inexorably destroying the last few examples of relatively undisturbed ecosystems that remain on earth, so that soon they will no longer be available for study. Brazilians are proclaiming that the Amazon rain forest will be gone in thirty years, to give one example, and no one can accurately predict what effects that massive change will have on South America and the world. Careful study is still needed.

The speed, scope, and intensity of interaction with the natural environment are crucially determined by the level of technology available to a human community. Using human and animal motive power and the energy of water, wind, and fire with the relatively simple tools and machines that had been invented, the ancient peoples constructed huge monuments which still impress us, but their level of interaction with the natural environment was relatively low as compared with that of modern industrial society. The changes wrought in the environment by ancient civilizations are massive indeed, but involved centuries or millennia for their accomplishment. Today more significant changes take place in months or years—or even seconds, in the case of atomic explosions. While the real extent and nature of the impact of ancient technology must not be underestimated, what impresses us almost as much is the failure of the ancients to pursue inventions, and the slow rate of technological change that resulted. Of all ancient peoples, the Romans possessed the most highly developed technology, and in this respect they are closest to us. Their machines for war, construction, and industry foreshadowed some that are still in use today. The fact that ancient peoples absorbed and survived successive changes in the technology of war and peace cannot be of much comfort to us today, because the rapidity, size, and power of such changes today are of an entirely different order from anything experienced in ancient times.

Another factor determining the way a human community will interact with the natural environment is the degree of organization and social control the community possesses. This is true because environmental ends desired for the good of the community may involve sacrifices on the part of its individual members, sacrifices which they would not make without some degree of social encouragement or coercion. The early civilizations of the river valleys, for example, had to be able to call for large expenditures of human energy on the construction of canals which seemed to benefit the entire society. Ancient civilizations were able to exert a considerable degree of social control because the vast majority of ancient people regarded themselves primarily as parts of their societies, and only secondarily, if at all, as individuals. Each person had a place in the social hierarchy which was rigidly defined and rarely changed. This was true of the pharaonic autocracy of Egypt, perhaps the most marked example of social control in the ancient world. But Egypt suffered periodic breakdowns in social control, and no ancient civilization could have channeled the actions of its citizens with regard to the environment to the extent that is at least theoretically possible today.

All ancient societies depended to a large extent on slave labor, a fact which seems to indicate an extreme degree of social control, until it is remembered that the majority of slaves were owned by citizens, not by the state, and that citizens were to a surprising degree able to pursue their private goals, at least in Greece and Rome.

As indicated in the preceding chapters, Greek and Roman governments established policies in the fields of agriculture, forestry, mining, and commerce, but citizens were allowed a wide latitude of choice within certain guidelines. Greek citizens had carefully defined duties to the community, but the city-states are noted for the freedom they allowed. The later Roman Empire tried to interfere in and control the lives of its citizens to an unusual extent; the edicts of Diocletian attempted to stabilize occupations, regulate prices, and control religion, while his secret police kept him informed of activities dangerous to the state. But no ancient autocracy remotely approached the ability of a modern industrial state to keep informed about its citizens and see that they perform their social duties. Greece operated without imprisonment as a punishment, and the Roman Empire supported itself financially without an income tax. The degree of control that can be exercised in the modern world by governments with electronic surveillance, computers, chemical and psychological methods, bureaucracies, police, deportations, and prisons is unmatched by anything seen before in world history. In democracies, environmental policies can be established only when widespread public support for them exists. Over the last few decades, such policies have in part been established even over the opposition of powerful pressure groups. Some needed measures have been blocked by the same groups or by the tendency of the public to prefer short-term personal gains to long-term benefit for society. Totalitarian states such as the Soviet Union have also taken some steps to preserve the environment, but we have not yet seen a major government take all of the steps which seem called for in the present ecological dilemma. Neither in ancient times nor in modern times have human communities become fully aware of the role which their relationship to the natural environment plays in their long-term welfare and even survival.

One conclusion which seems clear to this author is that the modern ecological crisis grew out of roots which lie deep in the ancient world, particularly in Greece and Rome. The problems of human communities with the natural environment did not begin suddenly with the ecological awakening of the 1960s, nor indeed with the onset of the Industrial Revolution or the Christian Middle Ages. Mankind has been challenged to find a way of living with nature from the earliest times, and many of our habitual answers to that challenge received their first conscious formulation within ancient societies, especially the classical civilizations.

At this point, one might well ask whether this study of ancient civilizations has produced any insights which might be of use in meeting the present crisis. If our ecological crisis has ancient roots, it might also be possible to learn from some of the successes and failures of ancient civilizations as we look to the future.

First of all, it might be possible for people today to recover something like the attitude of respect for the earth and nature that was felt by many in ancient societies. This could come not as a renaissance of animism, or a revival of ancient religions which have lost their ability to infuse human minds, but as a new insight compatible with many religions and philosophies. Judaism and Christianity could expand their concepts of human stewardship to recapture the biblical inclusion of the whole created natural world within the responsibility of people to God.

Islam has its own unique insights along similar lines. Others will be impressed by Albert Schweitzer's demonstration that the concept of reverence for life serves as a basis for philosophical ethics. Eastern philosophies, which have long contained attitudes toward nature which emphasize harmony, respect, and refusal to exploit, might find ways to realize their insights. Recent interest in the American Indian feeling for the land and its creatures reveals that the native Americans had ecological wisdom which can be studied and emulated. Better attitudes toward the natural environment will have to develop in a pluralistic human community, as people of varying traditions and points of view come to see the necessity of caring for the earth in order to preserve life itself and improve the quality of life.

Second, a concentrated effort to study the natural environment in all of its facets and interrelationships is needed. This is particularly crucial at the present moment, before much of the evidence about nature is altered, marred, or erased by human activities. These activities themselves and their effects upon the natural environment must also be investigated thoroughly. No wise environmental policy can be based on ignorance of the workings of nature. So that we may learn from what mankind has experienced through millennia of interaction with nature, more research is needed into the ecological relationships of past human societies, to correct and fill out the broad outline which is presented here.

Third, each human community must seek a viable relationship with the natural environment at the level made possible by the technology available to it. A study of ancient civilizations should demonstrate that a rejection of modern technology or an attempt to turn back the clock would not in itself assure a proper balance with nature or prevent environmental degradation. Rather, we should find ways to use our technological abilities in order to minimize the destructive impact of our civilization upon the natural evironment and to enhance our relationship with nature in ways which are beneficial both to people and to the environment. This would no doubt mean that some possible avenues of technological development ought to be abandoned, and that human population ought to be stabilized at some optimum size. No level of technology could support an unlimited increase of human numbers without catastrophic damage to the natural world and resultant crisis for mankind.

Finally, as human beings, we must be willing to accept freely certain limitations on our actions which affect the earth. In democracies, these limitations can be based on public awareness of the magnitude of environmental problems and of the options which exist to meet them. The alternatives to freely chosen environmental policies, consistently administered, are probably few. History does not provide us with an example of an ecologically aware dictatorship, willing to coerce its people to take the courses of action which it deems necessary for survival in balance with nature, but such a government is certainly a future possibility somewhere in the world, unpleasant as it may be to contemplate. History does, however, provide us with many examples of ancient peoples who failed to adapt themselves to live in harmony with the ecosystems within which they found themselves, who depleted their environment, exhausted their resources, and exist today only as ruins within eroded and desiccated landscapes. That fate might also await our own civilization, but this time on a global scale. Ancient history is a warning and a challenge to our attitudes, our ability to understand, our technological competence, and our willingness to make far-reaching decisions. The challenge will not go away, and the response we will make is not yet clear.

Notes

Chapter 1

1. Plato *Critias* 111B–D.

2. Ellen Churchill Semple, *The Geography of the Mediterranean Region* (New York: Henry Holt, 1931), p. 100.

Chapter 2

1. Homer *Odyssey* 9. 82 ff.

2. Semple, *Geography,* p. 17.

3. Isa. 40:7–8.

Chapter 3

1. Matt. 25:24.

2. James Mellaart, *Çatal Hüyük: A Neolithic Town in Anatolia* (London: Thames & Hudson, 1967).

Chapter 4

1. G. M. Lees and N. L. Falcon, "The Geographical History of the Mesopotamian Plains," *The Geographical Journal* 118 (March 1952):24–39.

2. Isa. 18:1.

3. "The Hymn to the Aton," trans. John A. Wilson, in *Ancient Near Eastern Texts,* ed. James B. Pritchard (Princeton University Press, 1955), p. 370.

Chapter 5

1. Song of Sol. 2:11–13. This and most other biblical quotations herein are from the Revised Standard Version (New York, 1946–52).

2. Ps. 19:1.

3. Isa. 6:3.

4. Gen. 1:31.
5. Ps. 24:1.
6. Ps. 95:4–5.
7. Gen. 1:28.
8. Gen. 2:15.
9. Joel 1:10,12.
10. Joel 3:18.
11. Isa. 35:1.
12. 2 Chron. 26:10.
13. Joel 2:23.
14. "The Hymn to the Aton," p. 371.

Chapter 6

1. Homer *Odyssey* 14. 457–58.
2. Homer *Iliad* 21. 470–71.
3. Ibid., 8. 47.
4. Pausanias *Description of Greece* 8. 24. 7.
5. Ibid., 7. 22. 4.
6. Homer *Odyssey* 19. 592–93.
7. Homer *Iliad* 24. 54.
8. Ibid. (trans. A. T. Murray) 16. 384–92. Quotations from classical authors herein are taken, for the most part, from the Loeb Classical Library translations (Harvard-Heinemann).
9. Homer *Odyssey* 7. 132.
10. Ibid. (trans. A. T. Murray) 19. 109–14.

Chapter 7

1. Homer *Iliad* 2. 613, etc.; *Odyssey* 15. 295; *Iliad* 2. 757; 9. 151.
2. Homer *Iliad* 8. 555–59. Cf. 16. 297–300.
3. Homer *Odyssey* 5. 73–77.
4. Euripides *Danae,* fr. 316, in Augustus Nauck, *Tragicorum Graecorum Fragmenta* (Hildesheim: Georg Olms, 1964).
5. Aristotle *Politics* 1. 3. 7. 1256b20.
6. Sophocles *Antigone* 332–75.
7. Plato *Critias* 111B.
8. Herodotus *Histories* 3. 108; repeated in Plato *Protagoras* 321b.
9. Theophrastus *Metaphysics* 9 (34).
10. Theophrastus *Enquiry into Plants* 2. 2. 8.

Chapter 8

1. Plato *Critias* (trans. R. G. Bury) 111B–D.
2. Theophrastus *Enquiry into Plants* 5. 8. 1.; Strabo *Geography* 14. 6. 5.

Chapter 9

1. Ovid *Fasti* 3. 295–96.
2. Virgil *Georgics* 2. 173–74.
3. Pliny the Younger *Epistles* 1. 9. 6.
4. Cicero *On Friendship* 19. 68.
5. Virgil *Eclogues* 1. 82.
6. Virgil *Georgics* 2. 485.
7. Cicero *On the Nature of the Gods* (trans H. Rackham) 2. 13 (35).
8. Ibid., 2. 60 (152).
9. Ibid., 2. 34 (87).
10. Columella *On Agriculture* 1, pref. 1–3.

Chapter 10

1. Cato *On Agriculture* 139.
2. Lucretius *On the Nature of Things* 5. 1370–78.
3. S. Colum Gilfillan, "Roman Culture and Dysgenic Lead Poisoning," *Mankind Quarterly* 5 (January–March 1965):3–20.
4. Vitruvius *On Architecture* (trans. Frank Granger) 8. 11.
5. Strabo *Geography* 3. c.146.
6. Cicero *Cato the Elder on Old Age* 15.
7. Cicero *Republic* 2. 6.
8. Horace *Odes* 2. 15. 1–2.
9. Frontinus *Aqueducts of Rome* 1. 16.
10. Tacitus *Agricola* 30. 5.
11. Plutarch *Life of Caius Marius* (trans. Bernadotte Perrin) 21. 3.

Chapter 11

1. Pliny the Elder *Natural History* 12. 84.
2. Michael Rostovtzeff, *Social and Economic History of the Roman Empire,* 2d ed., 2 vols. (London: Oxford University Press, 1957), 2:197.
3. Cicero *On the Nature of the Gods* (trans. H. Rackham) 2. 60 (152).

Chapter 12

1. Matt. 6:28–29.
2. Mark 4:28.
3. Mark 6:31.
4. Acts 14:15.
5. 2 Cor. 9:10 (Isa. 55:10).
6. Matt. 6:26.
7. Luke 12:6.
8. Romans 1:20.
9. 1 Tim. 4:4.
10. Heb. 2:8.
11. Rom. 8:19–23.
12. John 1:2–3.
13. Heb. 1:3.
14. Col. 1:17, 20.
15. 1 Cor. 3:21–23.
16. 1 John 2:15.
17. 2 Cor. 4:18.
18. Heb. 1:11–12.
19. Rev. 21:5.

Chapter 13

1. Lynn White, "The Historical Roots of Our Ecologic Crisis," *Science* 155 (1967):1203–7.
2. Plato *Theaetetus* 160D (15).
3. 1 Cor. 3:19.

Suggestions for Further Reading

The bibliography that follows is a varied one, since the subject of this book verges upon several fields of knowledge. The list is, for the most part, limited to works that deal directly with the ancient world. This author is particularly indebted to a few outstanding studies which should be singled out for recommendation to the reader.

The pioneering study by George Perkins Marsh, *Man and Nature* (New York, 1864), is a classic, published in subsequent editions culminating in *The Earth as Modified by Human Action* (New York, 1885). Marsh's scope includes the whole earth and all periods of history, but his comments on environmental history in ancient times are well developed and contain many valuable insights which have stood the test of time. In the twentieth century, another work heralded and helped to bring about a new period of environmental concern. This was Paul Bigelow Sears's *Deserts on the March* (Norman: University of Oklahoma Press, 1935), which is centered on the American scene but gives good consideration to the ancient background.

A valuable, detailed study by Ellen Churchill Semple, *The Geography of the Mediterranean Region: Its Relation to Ancient History* (New York: Henry Holt, 1931), describes the Mediterranean ecosystem and most of the economic activities of the ancient peoples. A more recent work, *The Ancient Mediterranean* (New York: Charles Scribner's Sons, 1969) by Michael Grant, sets ancient history in its natural context in a most engaging way.

Two books specifically investigate ancient attitudes toward nature. Henry Rushton Fairclough's *Love of Nature among the Greeks and Romans* (New York: Longmans, Green, 1930) is a rewarding aesthetic study, and Clarence J. Glacken's monumental, wide-ranging *Traces on the Rhodian Shore* (Berkeley and Los Angeles: University of California Press, 1967) includes major sections on the Greeks and Romans, the Hellenistic Age, and Judeo-Christian thought. Glacken is concerned with the history of three ideas: nature as a cosmic plan, environmental determinism, and human alteration of the earth.

Books

Africa, Thomas W. *Science and the State in Greece and Rome.* Huntington, N.Y.: R. E. Krieger Publishing Co., 1968.

Allen, Katharine. *The Treatment of Nature in the Poetry of the Roman Republic.* Madison, Wis.: University of Wisconsin Press, 1899.

Ashby, Thomas. *The Roman Campagna in Classical Times.* 1927. Reprint. London: Benn, 1970.

Bailey, Cyril. *Phases in the Religion of Ancient Rome.* Berkeley: University of California Press, 1932.

Biese, Alfred. *Die Entwicklung des Naturgefuehls bei den Griechen und Roemern.* 2 vols. Kiel: Lipsius & Tischer, 1882–84.

Black, John. *The Dominion of Man: The Search for Ecological Responsibility.* Edinburgh: Edinburgh University Press, 1970.

Boak, Arthur E. R. *Manpower Shortage and the Fall of the Roman Empire in the West.* Ann Arbor, Mich.: University of Michigan Press, 1955.

Bolkestein, Hendrik. *Economic Life in Greece's Golden Age.* Leiden: E. J. Brill, 1958.

Bradford, John. *Ancient Landscapes in Europe and Asia.* London: G. Bell & Sons, 1957.

Brooks, C. E. P. *Climate through the Ages.* 1949. Reprint. New York: Dover Publications, 1970.

Burford, Alison. *Craftsmen in Greek and Roman Society.* Ithaca, N.Y.: Cornell University Press, 1972.

Butzer, Karl W. *Environment and Archaeology: An Ecological Approach to Prehistory.* Chicago: Aldine Publishing Co., 1964.

Carcopino, Jerome. *Daily Life in Ancient Rome.* New Haven, Conn.: Yale University Press, 1940.

Carrington, Richard. *The Mediterranean: Cradle of Western Culture.* New York: Viking Press, 1971.

Cary, Max. *The Geographic Background of Greek and Roman History.* London: Oxford University Press, 1949.

Claiborne, Robert. *Climate, Man, and History: An Irreverent View of the Human Environment.* New York: W. W. Norton & Co., 1970.

Clark, Grahame. *World Prehistory: A New Outline.* Cambridge, Eng.: Cambridge University Press, 1969.

Collingwood, R. G. *The Idea of Nature.* New York: Oxford University Press, 1945.

Collis, John Stewart. *The Triumph of the Tree.* New York: William Sloane Associates, 1954.

Cornwall, I. W. *The World of Ancient Man.* London: J. M. Dent & Sons, 1964.

Dale, Tom, and Carter, Vernon Gill. *Topsoil and Civilization.* Norman: University of Oklahoma Press, 1955.

Davies, Oliver. *Roman Mines in Europe.* Oxford: Oxford University Press, 1935.

De Camp, Lyon Sprague. *The Ancient Engineers.* Cambridge, Mass.: M.I.T. Press, 1970.

Dhalla, Maneckji Nusservanji. *Zoroastrian Civilization.* New York: Oxford University Press, 1922.

Dimbleby, G. W. *Plants and Archaeology.* London: John Baher Publishers, 1967.

Durrenberger, Robert W. *Environment and Man: A Bibliography.* Palo Alto, Calif.: National Press Books, 1970.
Evenari, Michael; Shanon, Leslie; and Tadmor, Naphtali. *The Negev.* Cambridge, Mass.: Harvard University Press, 1971.
Fairclough, Henry Rushton. *Attitude of the Greek Tragedians toward Nature.* Toronto: Rowsell, 1897.
Forbes, Robert James. *Studies in Ancient Technology.* 9 vols. Leiden: E. J. Brill, 1955–64.
Frank, Tenney. *An Economic History of Rome.* 1927. Reprint. New York: Gordon Press, 1962.
Frankfort, Henri, et al. *The Intellectual Adventure of Ancient Man: An Essay on Speculative Thought in the Ancient Near East.* Chicago: University of Chicago Press, 1946.
Friedlaender, Ludwig. *Roman Life and Manners under the Early Empire.* 4 vols. London: George Routledge & Sons, 1909–13.
Geikie, Archibald. *The Love of Nature among the Romans.* London: John Murray, 1912.
Glotz, Gustave. *Ancient Greece at Work.* 1920. Translation. New York: W. W. Norton & Co., 1967.
Glueck, Nelson. *Rivers in the Desert: A History of the Negev.* London, 1959. New York: W. W. Norton & Co., 1968.
Goulimis, Constantine M. *Wild Flowers of Greece.* Edited by W. T. Stearn. New York: Academic Press, 1969.
Greene, Edward Lee. *Landmarks of Botanical History.* Smithsonian Miscellaneous Collections, vol. 54. Washington, 1909.
Guggisberg, C. A. W. *Man and Wild Life.* London: Evans Brothers, 1970.
Guthrie, W. K. C. *In the Beginning: Some Greek Views on the Origins of Life and the Early State of Man.* Ithaca, N.Y.: Cornell University Press, 1957.
Hamerton, Philip Gilbert. *Landscape.* Boston: Roberts Brothers, 1885.
Harrison, Fairfax, ed. and trans. *Roman Farm Management.* New York: Macmillan, 1913.
Haverfield, Francis John. *Ancient Town-Planning.* Oxford: Clarendon Press, 1913.
Heichelheim, Fritz M. *An Ancient Economic History.* 3 vols. Leiden: A. W. Sijthoff's Uitgeversmaatschappij N.V., 1958–70.
Hodges, Henry. *Technology in the Ancient World.* London: Allen Lane, 1970.
Houston, James Macintosh. *The Western Mediterranean World.* London: Longmans, Green and Co., 1964.
Hyams, Edward. *Soil and Civilization.* London: Thames & Hudson, 1952.
Isaac, Erich. *The Geography of Domestication.* Englewood Cliffs, N.J.: Prentice-Hall, 1970.
Kees, Hermann. *Ancient Egypt: A Cultural Topography.* Chicago: University of Chicago Press, 1961.
Lanciani, Rodolfo Amadeo. *Ancient and Modern Rome.* 1925. Reprint. New York:

Cooper Square Publishers, 1963.
Lee, Norman E. *Harvests and Harvesting through the Ages.* Cambridge, Eng.: Cambridge University Press, 1960.
Lenz, Harald Othmar. *Botanik der alten Griechen und Roemer, deutsch in Auszuegen aus deren Schriften.* 1859. Reprint. Wiesbaden: Dr. Martin Saendig, 1966.
Levy, Jean-Philippe. *The Economic Life of the Ancient World.* Chicago: University of Chicago Press, 1964.
Lord, Russell, *The Care of the Earth.* New York: Thomas Nelson & Sons, 1962.
Louis, Paul. *Ancient Rome at Work.* New York: Alfred A. Knopf, 1927.
Ludwig, Emil. *The Mediterranean: Saga of a Sea.* New York: McGraw-Hill, 1942.
Mellaart, James, *Çatal Hüyük: A Neolithic Town in Anatolia.* London: Thames & Hudson, 1967.
———. *Earliest Civilizations of the Near East.* New York: McGraw-Hill, 1965.
Michell, Humfrey. *The Economics of Ancient Greece.* Cambridge, Eng.: Cambridge University Press, 1940.
Mickey, Karl B. *Man and the Soil.* Chicago: International Harvester Co., 1945.
Moule, C. F. D. *Man and Nature in the New Testament.* London: Athlone Press, 1964.
Nairn, A. E. M. *Descriptive Paleoclimatology.* New York: John Wiley & Sons, 1961.
Nasr, Seyyed Hossein. *The Encounter of Man and Nature: The Spiritual Crisis of Modern Man.* New York: Fernhill House, 1968.
Newbigin, Marion Isabel. *The Mediterranean Lands.* London: Christophers, 1924.
Palgrave, Francis T. *Landscape in Poetry from Homer to Tennyson.* London: Macmillan, 1897.
Platner, Samuel Ball, and Ashby, Thomas. *A Topographical Dictionary of Ancient Rome.* London: Oxford University Press, 1929.
Polunin, Oleg, and Huxley, Anthony. *Flowers of the Mediterranean.* Boston: Houghton Mifflin Co., 1966.
Raikes, Robert. *Water, Weather and Prehistory.* London: John Baker Publishers, 1967.
Robertson, D. S. *Handbook of Greek and Roman Architecture.* Cambridge, Eng.: Cambridge University Press, 1954. Chap. 12, "Greek and Roman Town-Planning."
Roebuck, Carl, ed. *The Muses at Work: Arts, Crafts, and Professions in Ancient Greece and Rome.* Cambridge, Mass.: M.I.T. Press, 1969.
Rostovtzeff, Michael. *The Social and Economic History of the Hellenistic World.* 3 vols. London: Oxford University Press, 1941.
———. *The Social and Economic History of the Roman Empire.* 2d ed. 2 vols. London: Oxford University Press, 1957.
Roux, Georges. *Ancient Iraq.* London: George Allen & Unwin, 1964.
Sarton, George. *A History of Science.* 2 vols. Cambridge, Mass.: Harvard University Press, 1952–59.
Sauer, Carl Ortwin. *Land and Life.* Edited by John Leighly. Berkeley and Los Angeles: University of California Press, 1963

Schwanitz, Franz. *The Origin of Cultivated Plants.* Cambridge, Mass.: Harvard University Press, 1966.

Schwarzbach, Martin. *Climates of the Past.* Princeton: D. Van Nostrand Co., 1963.

Scully, Vincent. *The Earth, the Temple, and the Gods: Greek Sacred Architecture.* New Haven, Conn.: Yale University Press, 1962.

Sears, Paul Bigelow. *The Ecology of Man.* Eugene, Ore: Oregon State System of Higher Education, 1957.

———. *Life and Environment.* New York: Columbia University Press, 1939.

Shairp, John Campbell. *The Poetic Interpretation of Nature.* Boston: Houghton Mifflin, 1889.

Shepard, Paul. *Man in the Landscape.* New York: Alfred A. Knopf, 1967.

———, and McKinley, Daniel, eds. *The Subversive Science: Essays toward an Ecology of Man.* Boston: Houghton Mifflin Co., 1969.

Singer, Charles. *A History of Biology.* London: Abelard-Schuman, 1959.

———, et al., eds. *A History of Technology.* 5 vols. New York: Oxford University Press, 1954–58.

Sjoberg, Gideon, *The Preindustrial City: Past and Present.* New York: Free Press, 1960.

Soutar, George. *Nature in Greek Poetry.* London: Oxford University Press, 1939.

Stamp, Laurence Dudley, ed. *A History of Land Use in Arid Regions.* Paris: UNESCO, 1961.

Stanley, Daniel J. *The Mediterranean Sea.* Stroudsburg, Pa.: Dowden, Hutchison & Ross, 1972.

Taylor, Henry Osborn. *Greek Biology and Medicine.* 1922. Reprint. New York: Cooper Square Publishers.

Thomas, William L., Jr., ed. *Man's Role in Changing the Face of the Earth.* 2 vols. Chicago: University of Chicago Press, 1956.

Thompson, Dorothy Burr, and Griswold, Ralph E. *Garden Lore of Ancient Athens.* American School of Classical Studies at Athens, Excavations of the Athenian Agora Picture Books, no. 8. Princeton, N.J.: Institute for Advanced Studies, 1963.

Thompson, J. Oliver. *History of Ancient Geography.* Cambridge, Eng.: Cambridge University Press, 1948.

Toutain, Jules. *The Economic Life of the Ancient World.* London: Routledge & Kegan Paul, 1930.

Toynbee, Arnold J. *Hannibal's Legacy: The Hannibalic War's Effects on Roman Life.* 2 vols. London: Oxford University Press, 1965.

Van Deman, Esther Boise. *The Building of the Roman Aqueducts.* Carnegie Institution of Washington, no. 423. Washington, D.C. 1934.

Vita-Finzi, Claudio. *The Mediterranean Valleys: Geological Changes in Historical Times.* Cambridge, Eng.: Cambridge University Press, 1969.

Wallace-Hadrill, D. S. *The Greek Patristic View of Nature.* New York: Barnes & Noble, 1968.

Ward-Perkins, J. B. *Cities of Ancient Greece and Italy: Planning in Classical Antiquity.*

New York: George Braziller, 1974.

Watson, Richard A., and Watson, Patty Jo. *Man and Nature: An Anthropological Essay in Human Ecology.* New York: Harcourt Brace Jovanovich, 1969.

Wheeler, Mortimer. *Early India and Pakistan.* New York: Frederick A. Praeger, 1959.

White, K. D. *A Bibliography of Roman Agriculture.* Reading, Eng.: University of Reading, 1970.

———. *Roman Farming.* Ithaca, N.Y.: Cornell University Press, 1970.

Williams, George H. *Wilderness and Paradise in Christian Thought.* New York: Harper & Row, 1962.

Wilson, John A. *The Burden of Egypt (The Culture of Ancient Egypt).* Chicago: University of Chicago Press, 1951.

Wright, H. E., Jr., and Martin, P. S., eds. *Pleistocene Extinctions: The Search for a Cause.* New Haven, Conn.: Yale University Press, 1967.

Wycherley, R. E. *How the Greeks Built Cities.* Garden City, N.Y.: Doubleday & Co., 1962.

Zeuner, Frederick E. *A History of Domesticated Animals.* New York: Harper & Row, 1963.

Zimmern, Alfred, *The Greek Commonwealth.* 1911. Reprint. London: Oxford University Press, 1956.

Other Sources

Bernardi, Aurelio. "The Economic Problems of the Roman Empire." *Studia et Documenta Historiae et Iuris* 31 (1965):110–70.

Bromehead, Cyril Edward Nowell. "The Early History of Water Supply." *Classical Journal* 99 (1942):142–51, 183–96.

Clark, John Graham Desmond. "Water in Antiquity." *Antiquity* 18 (1944):1–15.

Dales, George F. "The Decline of the Harappans." *Scientific American* 214 (1966):92–100.

Dawkins, R. M. "The Semantics of Greek Names for Plants." *Journal of Hellenic Studies* 56 (1936):1–11.

Dilke, Oswald Ashton Wentworth, and Dilke, Margaret S. "Terracina and the Pomptine Marshes." *Greece and Rome,* 2d ser. 8 (1961).

Finley, Moses I. "Technical Innovation and Economic Progress in the Ancient World." *Economic History Review.* 2d ser. 18 (1965):29–45.

Flannery, Kent V. "The Ecology of Early Food Production in Mesopotamia." *Science* 147 (1965):1247–56.

Forster, E. S. "Trees and Plants in Herodotus." *Classical Review* 56 (1942):57–63.

Fussell, George Edwin. "Farming Systems of the Classical Era." *Technology and Culture* 8 (1967):16–44.

Gilfillan, S. Colum. "Roman Culture and Dysgenic Lead Poisoning." *Mankind Quarterly* 5 (1965):3–20.

Gow, Andrew Sydenham Farrer. "The Ancient Plough." *Journal of Hellenic Studies* 34 (1914):249–75.

Hartwell, Kathleen. "Nature in Theocritus." *Classical Journal* 17 (1922):181–90.

Holch, A. E. "The Ecology of Theophrastus." Unpublished MS. Denver, Colo.: University of Denver, 1958.

Horn, R. C. "The Attitude of the Greeks toward Natural Scenery." *Classical Journal* 11 (1916):302–18.

Hubbell, Harry M. "Ptolemy's Zoo." *Classical Journal* 31 (1935):68–76.

Hyde, W. W. "The Ancient Appreciation of Mountain Scenery." *Classical Journal* 11 (1915):70–84.

Jasny, Naum. "Competition amongst Grains in Classical Antiquity." *American Historical Review* 47 (1942):747–64.

Judson, Sheldon. "Erosion Rates near Rome, Italy." *Science* 160 (1968):1444–46.

Kenyon, Kathleen M. "Jericho." *Archaeology* 20 (1967):268–75.

Lees, G. M., and Falcon, N. L. "The Geographical History of the Mesopotamian Plains." *The Geographical Journal* 118 (1952):24–39.

Lowdermilk, Walter C. *Conquest of the Land through 7,000 Years.* U.S. Department of Agriculture Information Bulletin no. 99 (1941).

———. "Lessons from the Old World to the Americas in Land Use." In *Smithsonian Report for 1943,* pp. 413–27. Washington, D.C.: Smithsonian Institution, 1944.

Lowry, S. T. "The Classical Greek Theory of Natural Resource Economics." *Land Economics* 41 (1965):203–8.

Lukermann, F. "The Concept of Location in Classical Geography." *Annals of the Association of American Geographers* 51 (1961):194–210.

Olson, L. "Cato's Views on the Farmer's Obligation to the Land." *Agricultural History* 19 (1945):129–32.

Price, Derek J. de Solla. "The Antikythera Machine." *Scientific American* 200 (1959):60–67.

Reifenburg, A. "The Struggle between the Desert and the Sown." *Desert Research: Proceedings of the Symposium, Jerusalem, 1952* (1953):378–89.

Russell, Franklin. "The Road to Ur." *Horizon* 14, no. 3 (1972):90–103.

Russell, Richard J. "Climatic Change through the Ages." *Climate and Man* (U.S. Department of Agriculture, Yearbook of Agriculture, 1941), pp. 67–97.

Solecki, Ralph S. "Shanidar Cave." *Scientific American* 197 (1957):58–64.

Stevens, Charles Rolfe. "A Theoretical Study of Geographic Reality and Human Concepts." Master's thesis, University of Denver, 1968.

Stevens, Courtenay Edward. "Agriculture and Rural Life in the Later Empire." In *Cambridge Economic History of Europe,* 2d ed., 6 vols. (Cambridge, Eng.: Cambridge University Press, 1966), 1:92–124.

Usher, Abbott Payson. "Soil Fertility, Soil Exhaustion and Their Historical Significance." *Quarterly Journal of Economics* 37 (1923):385–411.

Vita-Finzi, Claudio. "Roman Dams in Tripolitania." *Antiquity* 35 (1961):77–95.

Waterbolk, H. J. "Food Production in Prehistoric Europe." *Science* 162

(1968):1093–1101.

Wertime, Theodore A. "Man's First Encounters with Metallurgy." *Science* 146 (1964):1257–67.

White, Lynn. "The Historical Roots of Our Ecologic Crisis." *Science* 155 (1967):1203–7.

Wright, H. E., Jr. "Natural Environment of Early Food Production North of Mesopotamia." *Science* 161 (1968):334–39.

Wulff, H. E. "The Qanats of Iran." *Scientific American* 218 (1968):94–105.

Yeo, Cedric A. "The Overgrazing of Ranch Lands in Ancient Italy." *Transactions and Proceedings of the American Philological Association* 79 (1948):275–307.

Index